Survey Automation
Report and Workshop Proceedings

Oversight Committee for the Workshop on Survey Automation

Daniel L. Cork, Michael L. Cohen, Robert Groves,
and William Kalsbeek, *Editors*

Committee on National Statistics
Division of Behavioral and Social Sciences and Education

NATIONAL RESEARCH COUNCIL
OF THE NATIONAL ACADEMIES

THE NATIONAL ACADEMIES PRESS
Washington, D.C.
www.nap.edu

THE NATIONAL ACADEMIES PRESS   500 Fifth Street, NW   Washington, DC 20001

The project that is the subject of this report was supported by contract no. SBR-9709489 between the National Academy of Sciences and the National Science Foundation. Any opinions, findings, conclusions, or recommendations expressed in this publication are those of the author(s) and do not necessarily reflect the views of the organizations or agencies that provided support for the project.

International Standard Book Number 0-309-08930-1 (book)
International Standard Book Number 0-309-51010-4 (PDF)
Library of Congress Control Number: 2003106250

Additional copies of this report are available from the National Academies Press, 500 Fifth Street, NW, Washington, D.C. 20001; (202) 334-3096; Internet, http://www.nap.edu

Suggested citation: National Research Council (2003). *Survey Automation: Report and Workshop Proceedings*. Oversight Committee for the Workshop on Survey Automation. Daniel L. Cork, Michael L. Cohen, Robert Groves, and William Kalsbeek, eds. Committee on National Statistics. Washington, DC: The National Academies Press.

OVERSIGHT COMMITTEE FOR THE WORKSHOP ON SURVEY AUTOMATION

ROBERT M. GROVES *(Co-Chair)*, Survey Research Center, University of Michigan, and Joint Program in Survey Methodology
WILLIAM KALSBEEK *(Co-Chair)*, Survey Research Unit, Department of Biostatistics, University of North Carolina
MICK P. COUPER, Survey Research Center, University of Michigan, and Joint Program in Survey Methodology
JOEL L. HOROWITZ, Department of Economics, Northwestern University
DARYL PREGIBON, AT&T Labs—Research, Florham Park, New Jersey

DANIEL L. CORK, *Study Director*
MICHAEL L. COHEN, *Senior Program Officer*
MICHAEL SIRI, *Program Assistant*

Preface

This volume on survey automation differs in structure from other workshop reports issued by the National Academies. We have chosen to present this finished volume as the combination of two sub-reports:

- *The proceedings of the workshop,* as it occurred on April 15–16, 2002. This is a transcript of the workshop presentations, edited for basic flow and to include such presentation graphics as are essential to effectively convey the points of the presentations.

- *A short report by the workshop's oversight committee*, containing the committee's reactions to the proceedings of the workshop and providing its recommendations.

These two reports—the report and the proceedings—are packaged together in this single volume to provide a unified discussion of the workshop material. We believe that putting the committee's conclusions in a concise report is an effective means of communicating those results, while packaging the short report with the proceedings provides all the relevant back-up and reference material. The report is Part I of the volume; the proceedings is Part II. The surrounding sections—such as the references and acknowledgments—have been constructed such as to be applicable to both sub-reports.

The text of this report contains references to particular company and trade names, including references to specific computer software packages. Such identification of specific names should not be interpreted as endorsement by the authoring committee or the National Academies, nor should it imply that the specific products are the best available for specific purposes.

Acknowledgments

The authoring committee for the Workshop on Survey Automation extends its thanks to the many people who made the workshop possible and whose contributions helped to bridge the computer science and survey methodology communities.

Our thanks first to the U.S. Census Bureau for its sponsorship of the workshop. Through a long and convoluted path from the project's initiation to the completion of the workshop, Pat Doyle and her staff provided much useful assistance and occasional prodding, and were most receptive to questions and suggestions.

The shape and content of the workshop took form rapidly after a very successful planning meeting on December 11, 2001—arranged at the request of the authoring committee—that brought selected computer scientists into the same room with the Census Bureau's practitioners. This planning session was most useful in clarifying paths of approach to the documentation and testing problems. Pat Doyle, Janis Lea Brown, and other Census staff put together a thorough briefing within a very rapid time frame. The authoring committee is most grateful to the three computer scientists called in for the meeting—Jesse Poore, Lawrence Markosian, and James Whittaker (Florida Institute of Technology)—for their eagerness to take on a new problem in their experience. Unfortunately, other commitments precluded Whittaker from participation in the workshop itself; we nonetheless appreciate his guidance at the early stages.

About the time the workshop took place, the staff was asked by the workshop's parent committee—the Committee on National Statistics (CNSTAT)—to present a synopsis of the workshop material at the public seminar portion of CNSTAT's regular meeting on May 8, 2002. We are very grateful to two workshop participants—Jesse Poore and Mark Pierzchala—for reprising their workshop presentations, on short notice, at the CNSTAT seminar.

Travel and logistics arrangements for the workshop were deftly made by the workshop's project assistant, Michael Siri. We also appreciate the last-minute help of Danelle Dessaint of the CNSTAT staff in arranging the participants' dinner.

Part I of this volume was reviewed in draft form by individuals chosen for their diverse perspectives and technical expertise, in accordance with procedures approved by the Report Review Committee of the National Research Council (NRC). The purpose of this independent review is to provide candid and critical comments that will assist the institution in making the published reports as sound as possible and to ensure that the reports meet institutional requirements for objectivity, evidence, and responsiveness to the study charge. The review comments and draft manuscript remain confidential to protect the integrity of the deliberative process.

We thank the following individuals for their participation in the review of Part I of this volume: Don A. Dillman, Departments of Sociology and Rural Sociology, Washington State University; William L. Nicholls II, consultant, Alexandria, Virginia; James O'Reilly, Blaise Services at Westat, Durham, North Carolina; Stacy Prowell, Software Quality Research Laboratory, University of Tennessee; Nora Cate Schaeffer, Department of Sociology, University of Wisconsin; Elizabeth Stephenson, Institute for Social Science Research, University of California at Los Angeles; and Dave Zubrow, Software Engineering Institute, Carnegie Mellon University.

Although the reviewers listed above provided many constructive comments and suggestions, they were not asked to endorse the conclusions or recommendations, nor did they see the final drafts of the reports before their release. The review of Part I was overseen by Richard Kulka, Social and Statistical Sciences, RTI International, Research Triangle Park, North Carolina. Appointed by the National Research Council, he was responsible for making certain that an independent examination of this report was carried out in accordance with institutional procedures and that all review comments were carefully considered. Responsibility for the final content of this report rests entirely with the authoring committee and the institution.

Robert Groves, *Co-Chair*
William Kalsbeek, *Co-Chair*
Workshop on Survey Automation

Contents

I Report   **1**

Introduction 5

Current Practice in Documentation and Testing 9

Shift from Survey Research to Software Engineering 12

Changing Survey Management Processes to Suit Software
    Design 14

Dealing with Complexity: Broadening the Concept of
    Documentation 22

Reducing Insularity 28

II Proceedings   **33**

Opening Remarks 35

What Makes the CAI Testing and Documentation Problems
    So Hard to Solve? *Pat Doyle* 36

Software Engineering—The Way to Be. *Jesse Poore* 63

Automation and Federal Statistical Surveys. *Bob Groves* 78

Understanding the Documentation Problem for Complex
    Census Bureau Computer Assisted Questionnaires.
    Thomas Piazza 83

The TADEQ Project: Documentation of Electronic
    Questionnaires. *Jelke Bethlehem* 97

Computer Science Approaches: Visualization Tools and
    Software Metrics. *Thomas McCabe* 116

Model-Based Testing in Survey Automation. *Harry Robinson* . . 137

Quality Right from the Start: The Methodology of Building
    Testing into the Product. *Robert Smith* 153

Interactive Survey Development: An Integrated View. *Lawrence
    Markosian* 164

Practitioner Needs and Reactions to Computer Science
    Approaches. *Mark Pierzchala* 174

Web-Based Data Collection. *Roger Tourangeau* 183

Interface of Survey Methods with Geographic Information
    Systems. *Sarah Nusser* 198

Prospects for Survey Data Collection Using Pen-Based
    Computers. *Jay Levinsohn and Martin Meyer* 211

Panel Discussion: How Can Computer Science and Survey
    Methodology Best Interact in the Future? 226

Workshop Information   **247**
  Agenda 247
  List of Workshop Participants and Attendees 249

Bibliography   **251**

Biographical Sketches of Workshop Participants and Staff   **253**

List of Figures

I-1   Example of paper-and-pencil-style questionnaire, as seen by an interviewer. . . 6
I-2   Example of questionnaire item flow patterns in a CAI instrument. 7
I-3   Prototype product line architecture for a CAPI process. 16
I-4   Effect on mathematical complexity of a small change in code. 26
II-1   One-page excerpt (out of 63) from the core questionnaire document, Wave 3
        of the Survey of Income and Program Participation (SIPP), 1993 Panel. . . . 38
II-2   General structure of a product line architecture. 66
II-3   Schematic model of a successful software environment. 69
II-4   Conjectured multipliers on cost of correcting errors at different phases of a
        software design project. 74
II-5   Item ME16 from Instrument Document (IDOC) for Wave 6 of the Survey of
        Income and Program Participation (SIPP). 87
II-6   "More Information About This Item" View of Item ME16 from Instrument
        Document (IDOC) for Wave 6 of the Survey of Income and Program
        Participation (SIPP). 88
II-7   Portion of a sample questionnaire, as it might be represented on paper. 99
II-8   Portion of a sample questionnaire, as it might be represented in CASES. . . . 100
II-9   Portion of a sample questionnaire, as it might be represented in Blaise. 102
II-10   Hypothetical routing graph of a questionnaire. 103
II-11   Sample question, as coded in the Questionnaire Definition Language used in
        the TADEQ project. 106
II-12   Sample route instruction, as coded in the Questionnaire Definition Language
        (QDL) used in the TADEQ project. 107
II-13   Screen shot of TADEQ applied to a sample questionnaire, with some
        sub-questionnaires unfolded. 109
II-14   Screen shot of some route statistics generated by TADEQ for a sample
        questionnaire. 111
II-15   Simple C algorithm with flow graph. 122
II-16   More complicated C algorithm with flow graph. 123
II-17   Schematic diagram of error-prone algorithm. 125
II-18   Four sources of unstructured logic in software programs. 127
II-19   Choice A in software metrics quiz. 129
II-20   Choice B in software metrics quiz. 130
II-21   Example of large change in complexity that can be introduced by a single
        change in a software module. 131
II-22   Simple survey example. 141
II-23   Operational states of Windows clock application, viewed as a flow graph and
        a model. 142
II-24   Example of static design for a Web questionnaire. 190
II-25   Humanizing touches in a Web questionnaire interface. 191
II-26   Web instrument with human face added to personalize the instrument. . . . 193

List of Tables

II-1 Areas of Testing Within Computer-Assisted Survey Instruments 175
II-2 Response Rates from Explicit Mode Comparison Studies (List-Based Samples) 187

Part I

Report

For over 100 years, the evolution of modern survey methodology—using the theory of representative sampling to make inferences from a part of the population to the whole—has been paralleled by a drive toward automation, harnessing technology and computerization to make parts of the survey process easier, faster, or better. Early steps toward survey automation include the use of punch (Hollerith) cards in tabulating the 1890 decennial census. The collaboration of surveys and technology continued with the use of UNIVAC I to assist in processing the 1950 census and the use of computers to assist in imputation for nonresponse, among other developments. Beginning in the 1970s, computer-assisted interviewing (CAI) methods emerged as a particularly beneficial technological development in the survey world. Computer-assisted telephone interviewing (CATI) allows interviewers to administer a survey instrument via telephone and capture responses electronically. The availability of portable computers in the late 1980s ushered in computer-assisted personal interviewing (CAPI), in which interviewers administer a survey instrument to respondents using a computerized version of the questionnaire on a portable laptop computer.

CAI methods have proven to be extremely useful and beneficial in survey administration. However, as survey designers have come to depend on software-based questionnaires, some problems have become evident. Among these is the challenge of effectively documenting an electronic questionnaire: creating an understandable representation of the survey instrument so that users and survey analysts alike can follow the flow through a questionnaire's items and understand what information is being collected. Testing electronic questionnaires is also a major challenge, not only in the software sense of testing (e.g., ensuring that all possible paths through the questionnaire work correctly) but also in terms of such factors as usability, screen design, and wording. The problems of documentation and testing survey questionnaires are particularly acute for large and complicated instruments, such as those utilized in major federal surveys; for such large surveys, even relatively minor challenges in a questionnaire can produce lengthy delays in the fielding of surveys. The practical problems encountered in documentation and testing of CAI instruments suggest that this is an opportune time to reexamine not only the process of developing CAI instruments but also the future directions of survey automation writ large—for example, to see whether strategies for resolving current CAI problems provide guidance on how best to develop stand-alone surveys for administration via the Internet.

Accordingly, the Committee on National Statistics (CNSTAT) of the National Academies convened a Workshop on Survey Automation on April 15–16, 2002, with funding from the U.S. Census Bureau (by way of the National Science Foundation). The Workshop on Survey Automation brought together representatives from the survey research, computer science, and statistical communities. Presentations were designed so that survey researchers and methodologists could be shown directions for possible remedies to the problems of documentation and testing, and so that computer scientists could be introduced to the unique demands of survey research. The second day of the workshop suggested emerging technologies in survey research, each of which is an area in which further collaboration between survey researchers and methodologists and computer scientists could be fruitful.

The proceedings of the workshop—an edited transcript of the workshop presentations—are contained in this volume. In this short summary report, we draw from the proceedings and outline the major findings and themes that emerged from the workshop, among them:

- the need to retool survey management processes in order to facilitate incremental development and testing and to emphasize the use of development teams;

- the need to better integrate questionnaire documentation—and, more generally, the measure of complexity—into the instrument design process, making documentation a vital part throughout the development process rather than a post-production chore; and

- the need for the survey research community to reach beyond its walls for further expertise in computer science and related disciplines.

The issues and hence our comments in this report pertain to the entire computer-assisted survey research community—the collection of federal statistical agencies, independent survey organizations, software providers, and interested stakeholders engaged in modern survey methods. Understanding the challenges and crafting solutions will necessarily take communication and collaboration among government, industry, and academia.

At the outset, it is important for us to reiterate that the basis of this report is a one and one-half day workshop. The time limitation of a workshop affects the scope of material that can be covered. In structuring the workshop, we focused principally on CAPI issues and their implementation in large surveys, drawing on Census Bureau experience with CAPI in particular. In this context, documentation and testing problems—and the need for solutions—are likely to be acute. Although this report refers primarily to CAPI, a great deal of the discussion is applicable to computer-assisted interviewing in general. That said, this report of a workshop is necessarily more limited in scope than a complete study of automation in surveys would be.[1] Moreover, no single, short workshop presentation is sufficient to support strong recommendations in favor of any particular methodology or software package, and no such endorsement should be inferred from these remarks. As we discuss in greater detail later, more concrete guidance on specific approaches would require further collaboration—making survey researchers and computer scientists familiar with each other's work—of a scope beyond that of a single-shot overview workshop. Accordingly, we hope that the comments in this report are interpreted as suggestive rather than strictly prescriptive.

INTRODUCTION

Since the late 1980s, the advent of portable laptop computers has offered survey practitioners newfound opportunities and challenges. The new practice of computer-assisted personal interviewing (CAPI) inherited some features from the existing technology of computer-assisted telephone interviewing (CATI) but differed principally in the manner of administration. CATI interviewers typically operate out of fixed centers and are under more direct supervisory control. In contrast, CAPI interviewers are deployed in the field, thus enabling direct face-to-face contact between interviewer and respondent; the field nature of CAPI work also promised to give CAPI interviewers a greater measure of autonomy than their CATI counterparts. Furthermore, face-to-face CAPI questionnaires are often longer and more extensive than telephone surveys.

The great promise of computer-assisted interviewing (CAI) has always been its infinite potential for customization. By virtue of its electronic form, a CAI questionnaire can be tailor-fit to each respondent; skip sequences can be constructed to route respondents through only those questions that are applicable to them, based on their preceding answers. Drawing on previously entered data, the very wording of questions that appear on the laptop screen can be altered to reduce burden on the interviewer and to best fit the respondent. Such custom wordings may include references to "his" or "her" (based on an earlier answer to gender) rather than generic labels, or automatically computed reference periods based on the current date.

For example, Figure I-1 shows an excerpt from a paper version of the National Crime Victimization Survey. This excerpt is not meant as an

---

[1] In particular, it is important to note that there is a substantial literature related to the challenges of instrument development—as they have been experienced in the evolution of CATI (see, e.g., House, 1985) and in CAI more generally (Couper et al., 1998)—that we do not explicitly review here, given our mandate to report on the workshop.

HOUSEHOLD RESPONDENT'S VANDALISM SCREEN QUESTIONS

46a. Now I'd like to ask about ALL acts of vandalism that may have been committed during the last 6 months against YOUR household. Vandalism is the deliberate, intentional damage to or destruction of household property. Examples are breaking windows, slashing tires, and painting graffiti on walls.

Since ________ _______, 20 ____, has anyone intentionally damaged or destroyed property owned by you or someone else in your household?

(EXCLUDE any damage done in conjunction with incidents already mentioned.)

|557| 1 ☐ Yes
2 ☐ No – **SKIP** to Check Item G

46b. What kind of property was damaged or destroyed in this/these act(s) of vandalism? Anything else?

Continue asking "Anything else?" until you get a "No" response.

Mark (X) all property that was damaged or destroyed by vandalism during reference period.

|558| 1 ☐ Motor vehicle (including parts)
*   2 ☐ Bicycle (including parts)
3 ☐ Mailbox
4 ☐ House window/screen/door
5 ☐ Yard or garden (trees, shrubs, fence, etc.)
6 ☐ Furniture, other household goods
7 ☐ Clothing
8 ☐ Animal (pet, livestock, etc.)
9 ☐ Other – *Specify* ↙

46c. What kind of damage was done in this/these act(s) of vandalism? Anything else?

Continue asking "Anything else?" until you get a "No" response.

Mark (X) all kinds of damage by vandals that occurred during reference period.

|559| 1 ☐ Broken glass: window, windshield, glass in door, mirror
*   2 ☐ Defaced: marred, graffiti, dirtied
3 ☐ Burned: use of fire, heat or explosives
4 ☐ Drove into or ran over with vehicle
5 ☐ Other breaking or tearing
6 ☐ Injured or killed animals
7 ☐ Other – *Specify* ↙

46d. What was the total dollar amount of the damage caused by this/these act(s) of vandalism during the last 6 months?

|560| $ |00| – **SKIP** to Check

Figure I-1   Example of paper-and-pencil-style questionnaire, as seen by an interviewer.

NOTE: This an excerpt from a version of the National Crime Victimization Survey (NCVS), as it would be seen by an interviewer administering the questionnaire to a respondent (the NCVS is partially conducted by CATI).

exemplar of paper questionnaires; indeed, the particular layout used here may be considered old style, and cognitive research has suggested better ways to structure paper questions in order to guide flow. But the excerpt suggests the basic structure of a paper questionnaire. Here, a question asks the respondent if he or she has recently been the victim of vandalism; if not, questioning is supposed to jump to an entirely different set of questions. But if they have, questioning continues within the set of vandalism questions, asking what kind of property was damaged.

Figure I-2 illustrates how this brief questionnaire segment might be handled in a computerized survey instrument. The specific pieces of logic in Figure I-2 are drawn from an instrument document (IDOC); it differs from the actual CAI code in that it has been post-processed and the pointers to and from immediate next steps have been cleaned and

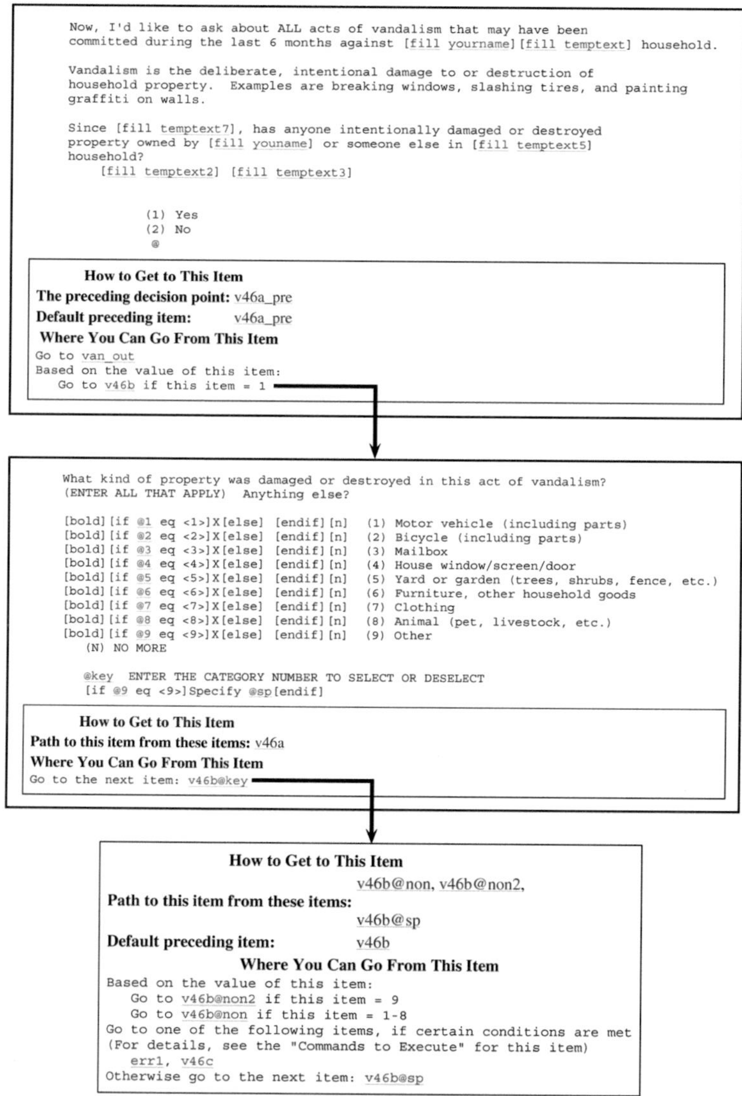

Figure I-2 Example of questionnaire item flow patterns in a CAI instrument.

NOTE: These excerpts parallel Figure I-1 in that they derive from the same portion of the National Crime Victimization Survey. However, the views shown here are those would be visible to a survey analyst, not an interviewer or a respondent; further, these are not portions of the raw computer code of the computerized questionnaire. These segments are excerpted from an instrument document (IDOC), the output of a specific automated documentation program on a CASES-coded version of the survey instrument.

formatted. Items referenced as [fill ⋯] are automatic fills, such as the reference date, and are filled in by the computer during the administration. The arrows and the "How to Get to This Item" boxes give a quick glimpse at the basic skip sequence governing this section of the questionnaire; the "Yes"/1 answer in the first question causes the jump to the second question on property type; and a response to that question causes a jump to a logical decision point (the lowermost box in Figure I-2), which will suggest a different path if the answer is "Other" than if the respondent gives one of the eight defined property types.

The use of CAPI also brought with it the promise of more accurate survey data. Errors caused by interviewers mistakenly skipping portions of paper questionnaires or by respondents being asked questions not applicable to their particular circumstances could be curbed by effective routing through the instrument. Computerization enables in-line editing and error checking; input values can be checked for accuracy and consistency (e.g., by asking for both age and birthdate and comparing the results); and flags can be raised, prompting the interviewer to solicit corrective information. Moreover, computerized questionnaires can facilitate easier "dependent interviewing" in longitudinal panel surveys, a practice in which answers from a respondent's questionnaire in a previous administration are used to frame questions on a current survey. Dependent interviewing can jog the memories of respondents in longitudinal surveys and provide for consistency in answers from wave to wave. Finally, data capture in CAPI surveys is automatic; answers need not be transcribed from paper forms or input using technologies like optical scanning.

The benefits of CAPI implementation—along with the experience of many successful conversions to CAPI—have led survey developers to pursue more extensive and complicated computerized questionnaires. The success of CAPI has brought with it a need and desire for added complexity. But the problem with an infinitely customizable instrument is that all the logical components therein—all of the potentially millions of logical paths through an instrument—must flow smoothly, because it is impossible to know ahead of time what specific path a particular respondent's answers may follow. All behind-the-scenes fills and calculations must operate properly to make sure that the questions are displayed on the screen correctly; all data input by the interviewer must be processed and coded correctly on the data output file to be of any utility. Thus, every computerized survey instrument must be a correctly functioning, error-free piece of software—a goal that is difficult enough for small surveys but greatly compounded by the sheer size and complexity of some of the federal surveys for which CAPI has been implemented.

The survey interview context adds another formidable challenge to constructing a quality computerized instrument. The skip sequences through an instrument may be thought of as prescribed flows, or forward motion through the questionnaire. But to truly accommodate the survey experience, a CAPI instrument must also have the capacity for unprescribed flows. If a respondent suddenly remembers income of $5,000 rather than the $1,000 reported earlier in the interview, there must be the facility to backtrack in the instrument, fill in a new answer, and proceed with the interview—either by returning to the previous spot in the instrument or by computing a new path based on the changed answer. Likewise, the instrument must be able to handle breaks at any time—cases in which the respondent opts not to answer some questions or to terminate the interview either temporarily or permanently.

CURRENT PRACTICE IN DOCUMENTATION AND TESTING

Two major challenges in conducting a survey through computer-assisted methods are the documentation and testing of computerized survey instruments. Documentation and testing are different in intent but strongly related in key ways. The strongest common link between the two may be that both are often thought of as end-of-the-line processes, things done when an instrument is complete and ready to be fielded (and done to the extent that remaining resources permit). However, documentation and testing are crucial parts of the questionnaire development process and fundamental to the quality and usability of the resulting data.

Documentation

Conversion from paper-and-pencil interviewing to computer-assisted methods effectively did away with a basic form of survey documentation—namely, a paper questionnaire that can be leafed through to see questions exactly as they would be worded when administered to respondents. A "free good" of paper-and-pencil survey development, the paper questionnaire itself is a weak form of documentation. Its usefulness in developing an understanding of the logical flows through sets of questions is limited, and it is not a guide or codebook for the data resulting from the survey. Still, paper has the major advantage of being tangible and therefore approachable to human reviewers in a way that software code to implement a survey is not. There is also a sense in which the scope of surveys—the number of conditional question sets that respondents of a certain type would get routed through—was kept in check by the medium of the paper questionnaire. Lest the document grow too massive and intimidate interviewer and respondent alike, survey instruments developed in the age of paper had a tendency to be somewhat shorter and simpler.

The relative unapproachability of computer code, the enabling of customized paths through a questionnaire, and the sheer magnitude of the extant questionnaires being converted by federal statistical agencies for CAPI implementation all combine to make the documentation of CAPI instruments a major problem. As difficult as it is to make sense of a paper questionnaire containing thousands of items, interpreting a computerized version to try to determine what exact questions are being asked and in what order is vastly harder. Moreover, for federal surveys, some manner of documentation that permits a gauge (of whatever accuracy) of respondent burden is a legal requirement, given the U.S. Office of Management and Budget's (OMB) statutory role in approving all government data collections.

Before a computerized survey instrument can be coded, specifications must be constructed so that programmers know what it is they are supposed to implement. These specifications, if kept and maintained as a living document over the course of the survey design process, could be a useful piece of documentation. In practice, as related at the workshop, current survey specifications are often in a perpetual state of development and difficult to keep current; for a federal survey, in particular, major changes in a survey may be called for on the fly in legislation or other agency interactions, further complicating the ability to maintain specifications. The Census Bureau reports a recent move toward using database management systems to develop and track specifications, although electronic specifications have proven to be as difficult to keep in synchronization as scattered paper specifications.

Documentation that outlines the inner logic of a questionnaire could also be a critical tool in the design process of a questionnaire. End users and OMB could benefit from documentation that suggests how specific survey items map to specific data needs and output data locations. Likewise, it could help survey designers map survey specifications to their implementations in the code and allow them to detect coding errors that may make it impossible for respondents to reach certain parts of a questionnaire. A subtle form of documentation that must also be considered in the computer-assisted interviewing arena is the archiving of computerized instruments, not only for potential reuse but also as a record of how specific surveys are conducted in an ever-changing computing environment.

Existing Systems: CASES/IDOC and Blaise/TADEQ

Methods to ease the creation of documentation by generating some form of it automatically from the electronic questionnaire itself have been prized goals. The computer-assisted survey community has made significant inroads in addressing the issue of automated instrument documentation. With sponsorship from the Census Bureau, the Computer Survey Methodology group at the University of California at Berkeley developed companion software for CASES, the DOS-based instrument authoring language that was a major force in early CAPI implementations but that has declined in use due to the lack of a Windows version. This companion software processes an instrument to produce an instrument document (IDOC), an automatically generated set of linked HTML pages that allow a user to browse through the logic of a questionnaire, identifying questions or decision points that flow directly into or out of a particular item in the questionnaire.

The emerging dominant survey authoring language, Blaise, also has a companion software suite for automated documentation under development. Sponsored by a consortium of European statistical agencies, the TADEQ Project has developed prototype software that can process a questionnaire and produce a flowchart-style overview of a questionnaire's logic, as well as some descriptive statistics. The eventual hope is for TADEQ to be independent of the software platform used to write the questionnaire—if an electronic questionnaire could be ported into the XML markup language, it could be processed by TADEQ—but initial development appears to have focused on its coordination with Blaise.

These two automated documentation initiatives are good first steps in addressing the global documentation problem in electronic surveys. Both suffer from some inherent practical limitations—IDOC from its exclusive applicability only to CASES-coded instruments and from the lack of an overall map to what can be a massive number of linked HTML pages, TADEQ from its perceived difficulty in processing very large and complicated instruments. More fundamentally, both systems are essentially post-processors of coded instruments; hence, the extent to which they can contribute to up-front guidance on questionnaire development—as a diagnostic tool during survey design—is not clear. Both also suffer from the reality that automated documentation can go only so far in conveying meaning and context to a human reader; it can suggest the functional flow from item to item but, on its own, it may not explain why those items are related to each other. Contextual tags and explanatory text in survey questionnaires require human input during coding (which often is not done, given time and resource limitations).

Testing

Less progress has been made in the area of comprehensive testing of CAPI instruments. Typically saved until the end of the design process, testing of CAPI instruments is—in the Census Bureau's experience—often tightly constrained by scant available time before due dates of deliverables and the remaining time and staff resources.

Current practice in testing CAPI instruments—the protocols used to test not only software functionality but also usability and human interface factors—differs across survey organizations. As indicated at the workshop, the Census Bureau relies to a large extent on the manual entry of a relatively limited number of sample answer scenarios. Some amount of direct manual testing of CAPI instruments is clearly necessary; cognitive factors, interviewer usability, and even mundane design choices like font size and screen color are things that can be properly appreciated only by directly interacting with the instrument. But hand input of dozens or even hundreds of prepackaged scenarios cannot possibly certify that all logical paths through a questionnaire work correctly. Manual scenarios must also be adjusted as the content of the questionnaire changes, making the process even more inefficient.

Part of the problem faced by the Census Bureau and other survey organizations in establishing a more rigorous test regime of a computerized survey's functionality lies in the limitations of the current software used to produce CAPI instruments. Neither CASES nor Blaise currently has the capacity for automated entry of randomly or probabilistically generated data to test paths through the instrument. At the workshop, it was mentioned that some prototype capacity has been developed to write scripts that can at least automate the task of entering data from canned answer scenarios into a Blaise-coded instrument, but automated test routines are not yet available.

SHIFT FROM SURVEY RESEARCH TO SOFTWARE ENGINEERING

Although the presentations at the Workshop on Survey Automation covered diverse topic areas, several unifying themes do emerge from reviewing the workshop proceedings. One of these is perhaps most fundamental: the advent of computer-assisted interviewing thrust survey research organizations and federal statistical agencies into the business of software development, a business for which the extant management styles and design processes of the survey world are ill suited. The major problems in documenting and testing computerized survey instruments are symptomatic of the need to reexamine the basic process of designing, constructing, and implementing a survey in the computer age.

This is not to say that the process of fielding surveys has become strictly an exercise in developing software. The root of a survey remains a communication process between interviewer and respondent, and the unique features of that communication interface—the sensitivity of respondents to factors like question wording and ordering, the need for interviewers to be able to convey questions with proper grammar in a language that the respondent can understand, and so forth—prevent survey research from being a truly mechanical science. Moreover, the translation from survey work to the arena of software development is not entirely direct. The number of end users of a commercially developed software package may number in the millions, dwarfing the total number of interviewers and respondents who encounter a typical CAPI implementation. Likewise, the number of iterations of a survey is usually quite small, while the latitude for continuous revision—reissues, updates, and patches—is greater in commercial software environments. But the uniqueness of the survey context does not give license to conclude that the documentation and testing problems in CAPI are so unique or so challenging that external expertise in software development cannot help address the problems.

It is perhaps telling that the methods currently in use have been termed "computer-assisted interviewing" or "computer-assisted survey information collection." These labels convey an implicit supremacy of the interview over the computer, of the end over the means. To a large extent, this implied supremacy is altogether appropriate; the credibility of survey research organizations and federal statistical agencies rests on the timely and accurate production of data, and highest priority is understandably put on the direct interface with respondents and on the final data products.

But an underlying message of the Workshop on Survey Automation is that the survey industry needs to begin thinking more in terms of "interview-oriented computing"—of the mechanics of designing effective software for data elicitation, capture, and processing. The process by which a modern survey is designed and developed must also result in a quality piece of software, and so that process must ensure that proper software engineering standards are met. The importance of rethinking the mechanics and the organizational structure of modern survey design becomes increasingly important as CAPI and CATI systems become more dominant features of the survey world, and more important still as computer-assisted surveys are implemented on new platforms, such as the Internet or handheld computing devices.

Building from this theme, other conclusions from the workshop fall into three basic categories:

1. *Reconciling survey design with software engineering.* Best practices in software engineering suggest the critical importance of building testing into the design process and of sequencing work on parts of a larger product. In software engineering terms, the objective is to think of computerized survey design as a product development project and consequently to select an appropriate life-cycle model for the questionnaire and software development process. The workshop presentations suggested specific techniques and organizational styles that may be useful in a retooled survey design system.

2. *Measuring and dealing with complexity.* The workshop offered a glimpse at strategies for assessing the complexity of computer code—mathematical measures based on graph theory that identify the number of basis paths through computer code and the degree to which code is logically structured into coherent, separable subsections. Such strategies may be used to go beyond the traditional concept of documentation in the survey automation context and underscore the bedrock importance of documentation in the survey design cycle. Attention to the mathematical complexity of CAI code may also be the impetus for much-needed discussion of the complexity of the surveys themselves—that is, the extent to which extensive use of CAI features like fills and backtracking adds value and functionality to the final code and offsets the programming costs those features incur.

3. *Reducing the insularity of the survey community.* The human communication interface and other features specific to surveys make the challenges of survey automation unique, but not so unique that experiences and practices outside the survey world are irrelevant. The survey industry should work to foster continuing ties with other areas of expertise in order to get necessary help.

These three categories are used to structure the remainder of this report.

CHANGING SURVEY MANAGEMENT PROCESSES TO SUIT SOFTWARE DESIGN

Computer science–oriented presentations at the workshop ranged from a discussion of high-level software engineering principles to a detailed outline of a newly emerging model for structuring labor and assignments in a large software project. The presentations are suggestive of some general directions that may be useful, particularly in terms of rethinking the broad outline of the CAPI survey design process.

Establish an Architecture and Standards

In his remarks, Jesse Poore emphasized the importance of a "product line architecture" in structuring large, ongoing software projects. A product line architecture is a product development strategy, a plan under which the major elements of a larger product are identified. Some elements are cross-cutting across various parts of the design process, while others are more limited in scope. This architecture then becomes a living document; it can be reviewed over the years and altered as necessary, but it endures as an organizational guide.

The major benefits of thinking in terms of an overarching product line architecture are twofold. First, it begins to scale back the overall difficulty of the task by emphasizing its modular nature. By dividing a large and complex task into more easily approachable pieces, implementing changes in parts of the process can be eased. Second, it emphasizes standards and the common features of product releases. A product line architecture defines elements that are common to all product releases as well as permissible points of variation. Having resolved the task of producing a product (e.g., a cellular phone) into modular (and to some extent common) pieces, the task of constructing a new version of the product is one of making selected changes in some parts—but not necessarily all parts, nor all at once. Hence, a new cellular phone will build from a substantial base of established work, and development of new features will have been fairly contained and manageable.

Poore provides a prototype product line architecture for the production of a CAPI instrument; it is shown in Figure I-3. What this architecture illustrates is, for instance, that security is a concern that cross-cuts all the stages of instrument development but that should be separable, in the sense that security protocols could be a common feature to multiple surveys. Accordingly, security mechanisms could be developed separately (allowing, of course, for inputs and outputs in each stage of instrument development) and need not be reinvented for each questionnaire development project. Likewise, the mechanics of statistical operations embedded in survey code could be considered another cross-cutting yet separable layer. This hypothetical product line architecture is, as Poore notes, purely illustrative, and the degree to which it is correct or needs refinement awaits future work. But the idea of developing a product line architecture in the survey context has definite merit, chiefly because it would stimulate development of standards within individual survey organizations and across the survey community as a whole.

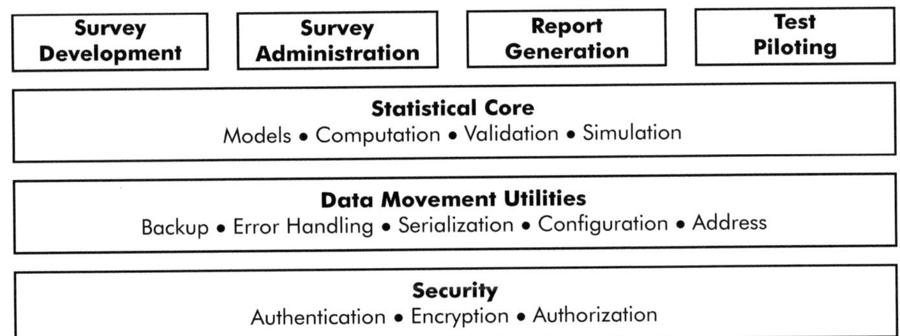

Figure I-3 Prototype product line architecture for a CAPI process.

NOTE: This figure is repeated later in the proceedings, in line with Jesse Poore's presentation.

SOURCE: Workshop presentation by Jesse Poore.

Although individual surveys vary in the specific content they cover, there is a similarity in functionality and structure—between different versions of the same survey and between different surveys entirely—that is not fully exploited in current CAPI implementation. To be clear, it is not survey measurement itself that we suggest needs standardization; individual organizations and researchers should always have wide latitude to define the information they hope to solicit from respondents and to craft questions accordingly. Instead, the need for standards arises because individual organizations find themselves reinventing even the most basic of structures—automatic fills of "his" or "her" based on an answer to gender, rosters in which names of household members can be stored and retrieved, questions of "yes/no" or "yes/no/other" types, for example. These mechanical structures are common across surveys within organizations as well as across entire survey organizations; they are instances in which industry standards or an archive of reusable code could ease programmers' burden and reduce errors.

Adopt Incremental Development and Testing and the Early Detection of Errors

During the first day of the workshop, a recurring topic was the difference in cost of software errors detected at the early stages of design and those found in later stages or after the software has been fielded. And although the bidding war that ensued throughout the day as to the exact magnitude of the multiplier on the cost of field-detected errors versus early detected errors—ranging from 32 to "astronomical"—added some levity to the day's proceedings, the underlying point is quite serious.

The formulation of a product line architecture is a first step used in contemporary computer science to try to detect errors early—to try to simplify the development process to reduce the opportunity for errors. A next step is to refine steps further and to develop work plans that are incremental in nature; this, too, is an organizational and behavioral change that could be useful in the survey context.

Incremental development is, as the name implies, a basic concept: resolve the large software/survey design task into modules or smaller pieces. These pieces can then be worked on separately and integrated one by one over time into the larger system under development. To borrow Poore's evocative phrasing, milestones (long-term deadlines and deliverable schedules) are always important to keep in mind, but it is also important to manage to what might be called "inch pebbles"—the parts of the project that will be done within a shorter time frame.

Testing is carried out during each stage of the integration process, rather than waiting until a large number of modules are simultaneously connected together at the final stage of development—the kind of en masse, last-minute, and depleted-budget activity that is unlikely to shed real light on the reliability of the finished product. Incremental tests or software inspections may also serve as a valuable feedback mechanism by pointing out defects in the specifications and design documents that are developed even before coding begins.

As Poore notes, an additional benefit of an incremental development strategy—beyond its integration of testing into the design process—is that it can help create an environment of success in the project team. Because of the difficulties of managing a complicated system as a whole, large projects can often founder; they can remain grounded at 70 or 80 percent complete for long periods of time when problems in one part bring the whole project to a standstill. In incremental development, there can be a greater sense of accomplishment; a particular module may be only a small part of the larger system, but there is value in being confident that this small piece is internally complete and ready for integration with the rest of the system (as well as further testing).

Together, an overarching product line architecture and incremental development may be particularly useful strategies in dealing with what was portrayed as a major problem in current survey work: requirements and specifications that are not well specified at the outset and that shift over time. This is particularly the case with surveys conducted by federal statistical agencies, for which new legislation or policy needs can suddenly shift the focus of an individual survey. The trick in rethinking the process is to construct an environment in which forward motion can always be made by proceeding with what is known and what is specified, and smaller parts of the project can be completed without being derailed by sudden changes elsewhere in the survey.

We certainly recognize that revising current survey practice to reflect these principles is no easy task. But identifying the set of problems currently experienced in CAI implementation and matching them to proposed solutions—whether through incorporation of best practices from software engineering or through industry-wide work on standardized solutions for the design of some CAI items—are important tasks, and ones that ought to be of considerable intellectual and professional excitement as well. It is, in short, a call for the survey industry as a whole to start a dialogue on standards and principles that have been lacking in the past.

On the topic of errors, a final practical point emerged at various times during the workshop. Although the aim is certainly to detect and fix errors early, there is no way to avoid errors entirely. Consequently, it is important for survey organizations to develop effective error tracking systems wherein errors at all stages of development—including those found in the field—can be logged and act as a feedback mechanism for designing the next iteration of a survey. It is possible that such an error tracking system could include building a "state capture" facility into CAI software, logging system events continuously through interviews and recording failure conditions when they occur. The log thus generated could then be very helpful in efficiently reproducing—and fixing—errors that caused the failure. It would also be beneficial to build error tracking tools in such a way that errors could be evaluated and classified by severity, considering both the cost of fixing them and the consequences for data quality of not fixing them. Defect tracking systems are common in software engineering usage, and hence discussion of the form of such a system for CAI implementations is an activity for which collaboration between the survey and software communities could be fruitful in the near term. Similarly, systems used in software engineering operations to configure and track design specifications could also be a useful short-term collaboration opportunity.

Employ Development Teams, Not Chains

Lawrence Markosian's presentation outlined one emerging framework—known as "extreme programming" (XP)—that is finding acceptance among software developers and takes incremental development as a basic principle. Intended to be applicable to high-risk systems in which specifications are often fluid, XP puts high emphasis on programming individual parts of code in pairs and frequent team meetings, emphasizing progress made on incremental parts of the project. Similar specific practices were also described during the course of the workshop: Poore advocated internal peer review of code and other work products, and Robert Smith's presentation described the general strategies used by the Computer Curriculum Corporation in structuring its software-building tasks.

As Smith commented in his remarks, it would be a mistake for the survey community to seize on any particular style for organizing software development and programming work—such as XP—and follow that style blindly. Instead, survey research organizations should be encouraged to research existing organizational styles and to experiment with them to the extent possible, ultimately forging a hybrid organizational strategy best suited to survey problems and context. Whether the specific mix of meetings and labor assignments in extreme programming is useful or even applicable in existing survey organizations—particularly the statistical agencies—is an open question, but such systems do revolve around central features that would be useful to consider.

The presentation of current survey practice at the workshop drew extensively from the Census Bureau's experience and, accordingly, descriptions of the labor organization in the survey process had a distinctly bureaucratic feel. Analysts (content matter specialists), designers, managers, programmers, cognitive testers—these and other roles in the survey design process were portrayed as being independent of each other and carrying out their jobs largely in isolation. The resulting structure seems to portray the survey-building process at the Census Bureau as strongly hierarchical and linear; the style seems to be unit-based, with responsibility for the entire survey being passed in turn from group to group and problems noted at one stage forcing the entire project to backtrack in the development process. Of course, other survey organizations—particularly nongovernmental entities—may structure the labor of questionnaire development differently. However, all organizational systems face important questions about responsibility for such cross-cutting concerns as documentation and testing: to what extent are survey managers, content matter specialists, or programmers responsible for instrument testing—responsible not only in terms of completing and evaluating the work but also budgeting time and resources for it to be done?

The concepts of extreme programming and peer review—as well as that of a product line architecture—suggest that it is desirable to consider task- or team-based structures. It may be best to have more input from programmers and field representatives early in the design process, and there may be benefits from investigating structures in which cognitive specialists are more members of the design team and perceived less as a level of oversight. Successful software engineering projects involve resolving the tasks into modular, addressable pieces and moving incremental pieces completely through the development cycle, rather than passing the entire project between task groups. It remains to structure the organization's labor strategy to work optimally under that framework, and survey organizations like the Census Bureau should be strongly encouraged to experiment with team-based programming styles.

Enable Automated Testing

In our view, attention to the general instrument development process could do much to correct some of the problems experienced in the CAPI implementation of large survey instruments. The extensive time delays that currently arise from very small changes in questions could be contained through a modular structure; attention to documentation and testing—including implementation of systems to achieve traceability and management of specifications throughout the development process— could dramatically improve the information available to developers and end users alike. Moreover, testing based on modular, incremental pieces of the project rather than a single end-of-project crunch should build confidence in much of the CAPI code. However, a basic truth remains that manual testing and hand input of selected scenarios are not adequate to certify that all of the myriad paths through a questionnaire will function properly.

As mentioned at the workshop, some facility has been developed to generate scripts that automate the steps of entering a test scenario for an instrument coded in Blaise. This manner of scripted testing is at the middle level of the trichotomy of test strategies suggested in Harry Robinson's presentation, the first level being manual testing. Developing capacity for these set preprogrammed scripts is an achievement but, as Robinson noted, automated test scripts are of quite limited use. As he put it, their shelf life is extremely short; virtually from the time they are developed they are out of date, since any change—particularly any change arising from a run-through of that selected script—can make the script useless. Scripts may make it possible to run more scenarios than would be possible under manual testing, but this still means that bugs can be detected only along the paths envisioned by the script and not elsewhere in the instrument.

The next tier of testing is automated testing, under which the computer makes random or probabilistically generated walks through the computer code. In the CAPI context, a simplified vision for how automated testing would work would be to put a range of possible answers on all the various questions, perhaps randomly drawn or perhaps based on previous survey administrations. An automated test routine would then step through the questionnaire, generating answers at each step along the way; by monitoring the progress, developers could find parts where the code breaks down. Moreover, automated testing could be used to assess the entire survey data collection process; the output dataset that emerges from some large number of tests could itself be analyzed to determine whether responses appear to be coded correctly. The practice of automated testing, of course, is harder than simplified visions allow, one in which active computer science research continues. Accordingly, automated testing is an area in which the survey community should monitor ongoing research to find techniques that suit their needs; the community then needs to work with their CAPI software providers to achieve the capability to implement those methods.

In his remarks, Robinson advocated one method for testing software systems that is growing in acceptance. The idea behind this method—called model-based testing—is to represent the functioning of a software system as various tours through a graph. The graph that is created for this purpose uses nodes to represent observable, user-relevant states (e.g., the performance of a computation, the opening of a file), and the arcs between a set of nodes represent the result of user-supplied actions or inputs (e.g., user-supplied answers to multiple-choice questions) that correspond to the functioning of the software in proceeding from one state of use to another. (A side benefit of developing a graphical model of the functioning of a software system is that it helps to identify ambiguities in the specifications.) The graphical representation of the software can be accomplished at varying levels of detail to focus attention on components of the system that are in need of more (or less) intensive testing. These graphical models are constructed at a very early stage in system development and are further developed in parallel with the system. Many users have found that the development of the graphical model is a useful source of documentation of the system and provides a helpful summary of its features.

Equipped with this model, automated tests can be set in motion. One could consider testing every path through the graph, but this would be prohibitively time-consuming for all but the simplest graphical models. Other strategies for which model-based testing algorithms have been applied include choosing test inputs in such a way that every arc between nodes in the graphical model is traversed at least once. Still other strategies include testing every path of less than n steps (for some integer n) or to conduct a purely random walk through the graph. Random walks are easy to implement and often useful, but they can be inefficient in terms of time because they may not cover a large graph quickly.

Markov chain usage models build on the graphical representation of a software program used in model-based testing. In this technique, a Markov chain probabilistic structure is affixed to the various user-supplied actions—arcs in the graphical model—that result in transitions from one node to the nodes that are linked to it. These (conditional) probabilities indicate which transitions from a given node are more or less likely based on the actions of a given type of user. These transition probabilities can govern a random drawing that is used to select paths through the graphical model for purposes of testing by selecting subsequent arcs, proceeding from one node to another. By selecting the test inputs in this way, the paths that are more frequently utilized by a user are chosen for testing with higher probability. Therefore, errors that are associated with the more frequent-use scenarios are more likely to be discovered and eliminated. An extremely important benefit of this testing is that long-run characteristics of system performance can be estimated based on well-understood properties of Markov chains. One such property is the expected length of the chain; by augmenting specific nodes with anticipated completion times and creating a continuous-time Markov chain, expected completion times with respect to all possible paths could be generated.

In his capstone discussion on the first day of the workshop, Mark Pierzchala listed three conclusions, with which we are in full agreement. First, the merits of model-based testing warrant further investigation by the survey community; it and other strategies for automated testing should be reviewed for their applicability. Second, implementing automated testing will be difficult given its novelty in the survey field; however, that difficulty does not outweigh the need for more rigorous testing than near-exclusive dependence on manual testing permits. Finally, a comprehensive CAPI test suite must always include checks of screen interface, usability, grammar, and other human interface features that are not amenable to automated testing.

DEALING WITH COMPLEXITY: BROADENING THE CONCEPT OF DOCUMENTATION

In designing the workshop, we asked Thomas Piazza and Jelke Bethlehem—the leading experts on the two existing (and evolving) systems for automated documentation of CAPI instruments, IDOC and TADEQ—to describe their programs and their approaches. To complement those talks, Thomas McCabe described research and tools to quantitatively measure the complexity of computer code and to visualize the internal logic of code. The IDOC and TADEQ talks centered on documentation in what is closer to the traditional sense of the word—a guide to the survey that can be handed or shown to users to give them an impression of what data are being collected and how specifically they are elicited. In comparison, McCabe's presentation is an invitation to view documentation more broadly, applying mathematical concepts of complexity in survey work, in ways that could ultimately become very important tools throughout the survey design process.

A first step in assessing complexity is to visualize computer code as a directed graph (flowgraph), wherein statements, decision points, and other pieces of code are designated as nodes and the links between them are edges (this graph of the functioning of the software is a much more code-dependent and detailed representation of the functioning of a software system than the graphical model that underlies model-based testing, described earlier). Having parsed the code as a graph, these flowgraphs may be plotted, and the resulting pictures may be extremely useful in directly seeing whether the logic underlying the code is clear or convoluted. More importantly, as McCabe has done, tools from mathematical graph theory can be applied to calculate natural metrics that objectively measure the complexity of the code.

At the workshop, McCabe focused on two such metrics. The first, cyclomatic complexity (v), is a measure of the logical complexity of a module (or a system). It can be computed by several methods, as illustrated in the proceedings, and it is defined as the number of basis paths in the graph. The set of basis paths is the minimum set of paths through the graph that—taken in combination with each other—can generate the complete set of individual paths. Accordingly, cyclomatic complexity is a benchmark of inherent complexity of the graph; it indicates the minimum number of separate tests that would be needed to cover all edges of the graph (or all the transitions in the software).

Essential complexity (ev) is a second metric that measures the degree of unstructuredness of software. Roughly speaking, essential complexity simplifies the flowgraph by fusing nodes that represent primitive (and well defined) logical structures; the collection of nodes that define a complete loop that executes while a certain condition is met may be replaced by a single node, for instance. Essential complexity is then calculated as the cyclomatic complexity of the resulting reduced graph. The quantity ev is bounded by 1 (perfectly structured, in that it can cleanly be resolved into well-defined logical modules) and v (perfectly unstructured, when no simplification is possible).

McCabe has developed many other metrics that could not be described at the workshop due to time constraints, among them a measure of the degree to which statements within modules make use of external data from other modules (thus indicating the success of the modularity of software design). Implemented for CAPI instruments, this metric would indicate how often edits within a given module make use of information collected in other modules.

There is great power in having an objective measure of complexity, and awareness of complexity throughout the survey/software design process could be a most useful diagnostic tool. Were it possible to conceive of a "first cut" measure of complexity that could be derived quickly and inexpensively from early specifications, those specifications and requirements could be reevaluated and solidified more quickly than is currently the case. Measuring the complexity of individual modules as the project moves along in increments would be a useful feedback mechanism, warning questionnaire designers of possible problem areas and urging the simplification of instrument routines. It is also reasonable to expect that complexity measures, in time, could prove to be a useful tool in budgeting and planning new CAPI instruments, since it may be possible to estimate the relationship between programmer/tester effort and expected levels of complexity.

Develop Prototype Complexity Measures for Existing Surveys

The software tools developed by McCabe and his former company act as parsers for a series of common languages, processing code and extracting the flowgraph and complexity measures. Of course, no premade software generates such measures for CASES or Blaise instruments, since those languages—being confined to the survey research arena—have a relatively small user base. However, it would be worthwhile for the survey research community to investigate the possibility of calculating McCabe-type metrics on existing CAPI instruments.

It should be noted that both of the existing automated documentation packages, IDOC and TADEQ, do generate what could be the raw materials for a complexity assessment, by resolving each questionnaire element into a node. Indeed, the prototype TADEQ software does offer the option to draw flowcharts of instruments. Trying to incorporate complexity measures into the existing TADEQ structure or converting CAPI instruments into an intermediate format that could be read by a McCabe-type parser could be the most promising ways to see how complexity measures work with existing survey code.

Develop Measures of Operational Complexity

Interpreting complexity as a mathematical construct offers a new way of thinking about and characterizing an instrument. But comments made at the workshop also suggest intriguing opportunities to build on the idea of complexity by constructing measures of what might be called "operational complexity."

Several workshop speakers and participants made remarks about the usefulness of trying to determine modal or characteristic paths through a questionnaire. From a practical standpoint, knowing that some high percentage of respondents tended to follow, say, 12 paths through an instrument would be a very valuable piece of information, as it could obviously set high priorities for testing. But knowing the nature of those paths—what questions or question patterns typify them—could also be of great value to end data users. So, too, could knowledge of those paths be valuable to OMB in its assessment of what questions a typical respondent will be administered and how long the survey will take. Research should be done on ways to best identify these modal paths, whether through testing (either pilot fieldwork or being derived from the output dataset from automated testing) or as a byproduct of a complexity assessment program.

On a related note, respondent burden—time spent in the interview—is of particular concern to OMB, which must approve federal surveys. According to discussion at the workshop, users have requested that the Blaise/TADEQ system allow coders the capacity to enter an estimated time necessary to complete a particular question; these times could then be added up for various paths by TADEQ to estimate modal and extreme completion times. Defining and enhancing such measures of respondent burden are also useful research concerns.

Although not mentioned at the workshop, other related measures of operational complexity may be useful to consider. For instance, backtracking in an instrument—respondents asking to go back and change a preceding answer—could be interpreted as a crude measure of confusion in an instrument. Accordingly, constructing post-processing systems that identify questions and modules that were most referenced in backtracking could be useful in future incremental development.

Complexity in the Code Versus Complexity in the Survey

Developing diagnostic measures may help manage the complexity that is due to automation of a survey, the complexity of the underlying computer code. However, coming to terms with complexity in modern surveys also requires serious thought about managing the complexity inherent to the survey itself.

The federal statistical agencies, in particular, are known for fielding survey instruments that are daunting in their scope and their complexity. As Pat Doyle said in her overview comments, the Survey of Income and Program Participation (SIPP) contains on the order of 13,000 items; some 200,000 items are covered in the Consumer Expenditure Survey (CES), the purchase data that are used in deriving the Consumer Price

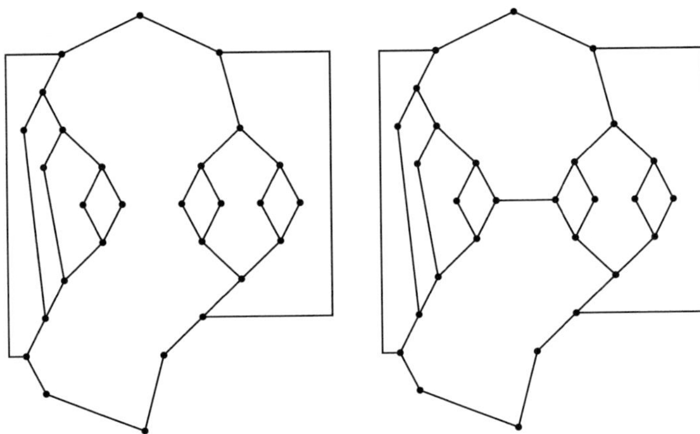

Figure I-4 Effect on mathematical complexity of a small change in code.

NOTE: This illustration is repeated in part later in the proceedings. Shown here are schematic diagrams with dots representing specific pieces of node—individual lines of code or subroutines—and the lines connecting them are function calls or links from routine to routine. The two diagrams differ by only one logical connection; however, the left graph has cyclomatic complexity 10 and essential complexity 1 (perfectly modular in structure), while the right graph has cyclomatic complexity 11 and essential complexity 10 (roughly, ten-elevenths logically confounded and hence largely incapable of modularization).

SOURCE: Workshop presentation by Thomas McCabe.

Index. Instruments like these are inherently complex because of their scope and public policy information needs, which dictate that high volumes of information be collected.

One lesson that follows from the discussion at the Workshop on Survey Automation is that there is a fundamental trade-off between having CAPI instruments that are infinitely customizable versus ones that are gratuitously complex. Paper instruments did not have the capacity for fills—for a question to recall a child's name or to refer to "his" or "her" on the basis of preceding answers—but CAPI instruments do. The degree to which large numbers of automated fills make a survey interview smoother, faster, or better is unknown and unproven, as is the effect of fills on ultimate data quality. Hence, the question is: Do fills unnecessarily complicate the underlying software? Does the cost (in both resources and complexity) of coding behind-the-scenes calculations to provide pristine wording of a question on a laptop screen outweigh that of allowing field interviewers to use their own judgment in reading a question using more generic labels? These cost-benefit trade-offs are important to consider. McCabe's illustrations suggest that mathematical complexity can be increased dramatically by even the smallest of changes in a piece of code. For example, Figure I-4 depicts a schematic piece of code for which adding a single link turns perfectly modularized code into a nearly intractable logical mess. By gaining a greater appreciation of the mechanical complexity of CAPI instruments, the survey industry may also be able to engage in an important discussion of whether the drive to automate surveys might, in some cases, cross the line of gratuitous complexity.

Another fundamental source of current coding complexity is the capacity for backtracking in a CAPI interview. To best simulate the paper-and-pencil survey experience, the standard that is hoped for in some quarters—complete ability to go back and revise and correct errors, in the same way that pages could be flipped back and forth—is crucial. But allowing backtracks obviously increases the complexity of the software task; these overriding "goto" statements can result in constructs that cause code to break down, such as jumps into loops and branching into decision nodes. Moreover, they raise concerns for the final output dataset. During the workshop discussion, it was not immediately clear what happens in current instruments when backtracks lead to the formation of a new path: Do the data already entered get voided automatically? Should those data be retained, in case the survey works itself back to the point at which the respondent asked to change an earlier answer?[2] Again, there is no apparent right or wrong answer, but the degree to which backward motion is permitted brings with it a cost-benefit trade-off in complexity.

To be clear, what we suggest in reexamining the complexity induced by mechanisms like fills and backtracking is an evaluation of trade-offs, not a single-minded drive to reduce complexity. Indeed, a move to absolutely minimize complexity could produce computerized survey instruments that may perform poorly in meeting survey demands or be unusable by interviewer. Moreover, it is not entirely clear that survey complexity is directly related to mathematical software complexity; that is, it is possible to imagine survey segments that seem intuitively complex but that could be quite simple software projects, and vice versa.

---

[2]Ultimately, it was concluded that Blaise and CASES both retain the data and have them available until the interview session is closed (Blaise) or a separate clean-up procedure is executed (CASES).

REDUCING INSULARITY

In making the preceding comments and recommendations, we do not in any way mean to imply that current survey design standards are "wrong" while software engineering standards are "right." As one of the workshop discussants aptly observed and McCabe echoed in his remarks, the software industry hardly has an unblemished record in providing error-free and perfectly modular code on budget and on time. As we stated earlier, there are no quick fixes in improving current CAPI implementation; there is no single software engineering panacea that can be applied.

That said, current practice in computer science offers ways of structuring software projects that could markedly improve CAPI implementation. The computer-assisted survey research community is a relatively small group who have pursued the first two decades of CAPI with great professionalism and curiosity, often making do with very limited resources. But the Workshop on Survey Automation suggests that furthering the CAPI cause will require solutions and approaches with which current survey practitioners may be unfamiliar. Accordingly, the survey world remains insular of developments in computer science and software engineering at its peril; opportunities for long-lasting collaboration should be actively pursued.

Survey research is a relatively small industry; as an activity within the federal government, the total budget for federal surveys is a very small pool. As a consequence, the burden for building bridges to other disciplines rests principally on the survey research community. Although the problems are interesting and formidable, the scale of survey computing and limited user base are such that computer scientists and software developers are unlikely to latch onto survey work as a viable work area of their own accord. It will take money and effort to build external connections (as well as to conduct vital information sharing and standard building within the survey community), but the benefit of outside experience is substantial. The task of building external connections to software expertise parallels the one that has been faced in other product development industries, such as consumer electronics. These industries typically begin developing software functionality using in-house resources, without reference to professional software engineers. Ultimately, dependence on software to deliver functionality has led companies to take a more professional approach to their software development activities.

The need for outreach to outside experts is increasingly important as time passes and technology advances. The second day of the workshop featured talks in three areas in which new and emerging technologies are beginning to become part of the survey experience:

- Development of surveys for deployment on the Internet (presented by Roger Tourangeau), thus removing a human interviewer from the process and requiring higher standards for human interface and usability;

- Incorporation of geographic information systems and global positioning satellite technology in the survey process (presented by Sarah Nusser), opening exciting new prospects for the development of survey frames and easing field interviewers' basic navigation and task work; and

- Migration of surveys from laptop computers to portable handheld computers (presented by Jay Levinsohn and Martin Meyer), literally lightening the burden of field interviewers while presenting new challenges in terms of reduced on-screen space and more limited battery capacity and storage space.

Coverage of these and other topics in general survey automation—among them the use of wireless networks and synchronization with the case management systems used to track completed questionnaires and assign follow-up interviews—was necessarily limited in this single workshop, and each topic merits fuller study.

Realizing the benefits of these and other new technologies will be difficult without increased attention to standards and practices in the survey industry or drawing on the expertise of fields outside of survey research. For example, one particular segment of contemporary survey work that suggests great potential challenges is the development of mixed-mode surveys. To boost response rates and improve survey coverage, survey designers increasingly consider conducting the same survey using multiple response modes (e.g., offering respondents the chance to reply either by mail or the Internet or conducting a mail survey but following up with nonrespondents via telephone). Thus, the inherent difficulties of implementing a survey in any particular medium are compounded by the problem of managing parallel versions of the same survey using different media. Assigning total responsibility for developing a survey in each response mode to different groups of workers seems an inefficient and possibly error-prone way to proceed, particularly if each group develops its own unique standards and processes to best suit their given response mode. Hence, in addition to reaching internal agreement on survey specifications and item types, incorporating a product line architecture (identifying and emphasizing common elements, such as data movement and processing routines) seems to be a vital step in making mixed-mode surveys work most effectively; so too is carefully weighing the trade-offs between needed functionality in one mode and added complexity thus incurred in another.

Foster Collaboration Beyond Walls of Survey Research

Survey research organizations and federal statistical agencies involved in survey work should seek ways to foster genuine collaboration with outside experts in computer science. Convening functions, such as the Workshop on Survey Automation or inviting computer science-based sessions at survey research meetings, are useful in this regard. However, creative ways to achieve true collaboration—engaging experts from both sides in detailed work, becoming familiar with each others' fields—are in greatest need.

A useful starting point to forge partnerships may lie in pilot work on some of the more immediately achievable recommendations suggested by this workshop. The task of finding a way to parse CAPI code in order to produce complexity metrics is one such project; another is selecting a nontrivial but still manageable existing survey and working out the mechanics of model-based testing for that instrument. Another immediate point of useful survey–computer science collaboration could be in tapping extant software engineering work on tracking project specifications over the life-cycle of a software project. The Census Bureau is working on tracking changes to survey specifications using database structures; this is an area in which outside expertise may be immediately applicable. As mentioned earlier in this report, drawing on software engineering best practices to design defect or error tracking systems is similarly an area in which collaboration could be fruitful.

Draw from Experience of Related Applications

At the workshop, Robert Smith spoke about the experiences of the Computer Curriculum Corporation in developing software for computer-based instruction. These software packages feature novel challenges for measuring advancement and deciding appropriate times to move students toward more advanced topics. But, at its heart, the course curriculum example shares many common features with CAPI: the question-and-answer format is central to both areas, both involve software projects of considerable size and scope, and both involve strong attention to human factors and communication issues.

The computer-assisted survey community should identify application areas like computer-based instruction packages that—while not exactly identical to CAPI implementation—share common features with the survey experience. These application areas may be existence proofs that problems similar to those in the CAPI world have been encountered by other practitioners, and those experiences may usefully be brought to bear in survey work.

In addition to computer curriculum software, another obvious analogue to CAPI survey collection is automated tax preparation software. Like a CAPI survey, tax software must be able to conform to the needs of a particular user using flow and skip sequences; like federal surveys in particular, the number of items a particular user may need to encounter will vary with their particular situations but could be quite large. In terms of production or release cycles, the annual revisions of tax software may be a better analogy to CAPI surveys than would other software projects. As a CAPI survey is to a respondent, tax software is also a case in which user exposure is fairly limited; for a particular user, the software effectively has one chance to work and it must work correctly. Indeed, the requirements of tax software may in some respects be more stringent than many CAPI instruments because tax software involves self-administration rather than a human intermediary; as traditional CAPI surveys evolve toward self-administered Internet-based questionnaires, the lessons learned in the tax preparation area could be particularly important.

Enhance Training of Current and Future Survey Researchers

Making changes as suggested in this report will require not only serious commitment by survey organizations but also a diffusion of new knowledge among existing survey development staff. Sustaining the changes will require that the skill sets of new survey practitioners reflect new organizational styles.

Accordingly, in her remarks, Pat Doyle stressed the importance of educating current and future survey staff about the importance and methods of documenting survey instruments, encouraging academic programs specializing in survey methodology to incorporate such training in their curricula. On this point, we agree; to the extent that software design is a key part of survey development, training in contemporary survey methodology should provide some background in software engineering. This includes not only effective strategies for instrument documentation, but also best practices in managing intensive software projects and emerging techniques for testing survey software as well.

Keep the Survey-Computer Science Discussion Active

In closing, it is our hope that the survey research industry will strive to build continuing channels of communication with the field of computer science. As several participants at the Workshop on Survey Automation commented, this workshop was not the first such gathering at which they had been in attendance. Over the past decades, several conferences on emerging technology and survey practice have been held, and they—like the Workshop on Survey Automation—serve a useful purpose.

Jesse Poore reminded the workshop participants of Moore's Law, the adage that technological capacity tends to double on a roughly 18-month cycle. In light of this observation, and of the great lessons that survey methodology and computer science have to offer each other, having joint activities like the Workshop on Survey Automation once a decade is clearly too infrequent to be useful. Regardless of the forum for such collaborations—whether workshops like this one, special sessions at professional meetings, or other means—the computer-assisted survey community should strive to have formal collaborative opportunities with computer science and related fields on at least a three-year (or twice Moore's Law) cycle. The greater challenge, as ever, will be to emerge from discussion and translate talk into action, forging enduring partnerships. Maintaining communication channels is too important a task to neglect.

Part II

Proceedings

OPENING REMARKS

CORK: Let me introduce myself. My name is Daniel Cork, and I am the study director for this, the [National Research Council (NRC)] Workshop on Survey Automation. This workshop is being conducted by the Committee on National Statistics of the NRC, with sponsorship from the U.S. Census Bureau. The agenda in the agenda books identifies Chet Bowie, who was going to give opening remarks on behalf of the Census Bureau; he is unable to make it, so actually we are going to have our first speaker—Pat Doyle, also of the Census Bureau—give those remarks in Chet Bowie's stead.

DOYLE: Welcome; I'm really thrilled that you all could come today and share with us your expertise on what we believe to be a very pressing set of issues. Basically, our task for the two days is to address our rather overzealous entry into computer-assisted personal interviewing (CAPI).[1] Automation is a wonderful thing; it allows us to basically take instruments to the point beyond which we can comprehend them. And we have certainly taken up that challenge.

We have instruments that provide a great deal of precision in measurement, and that's excellent for the quality and the statistics we can produce from our surveys.[2] It's also allowed us—when we try—to reduce the burden by targeting our questions precisely to the individuals [to] whom the questions would be relevant.

But all of this comes at the cost of complexity, and that complexity complicates our comprehension of the instrument. It complicates the testing of the instrument. It increases the time and the resources needed up front to get started. It prohibits an interpretable image of the instrument that's being fielded, i.e., the questionnaire. And basically it did away with the free good of instrument documentation. When we were in paper, we had a free good; we had documentation.

Now, we don't really believe [that] the testing challenges we face are new; we think they have been faced in other disciplines. And what

---

[1]Computer-assisted personal interviewing (CAPI) involves the administration of a computerized survey questionnaire by an interviewer to a respondent. The transition from traditional pen (or pencil)-and-paper interviewing (PAPI)—where the instrument is a paper document—to CAPI is the primary focus of the workshop. This focus differs slightly from related practices such as computer-assisted telephone interviewing (CATI); there, the questionnaire is also computerized, but CATI questionnaires tend to be shorter than CAPI questionnaires. Moreover, CATI interviewers are typically under direct supervisory control and operate from centralized call centers, whereas CAPI interviewers are deployed directly into the field to collect information and operate with somewhat greater autonomy. Generally, survey techniques involving electronic questionnaires are known as computer-assisted interviewing (CAI).

[2]In survey terminology, the questionnaire that is administered to the survey respondents is called an "instrument."

we hope to do as part of this workshop is to learn from those other disciplines how we can better do the testing of the instruments that we are creating.

We have considerable documentation challenges to face, and I believe a lot of them are going to require behavior modification as well as new tools. And some would say more management as well; since Chet's not here, I felt more comfortable in saying that one.

I think that it's important for those of you involved in the education of survey methodologists [to know] that we need to increase the education of both our current staff and our potential new staff in areas related to the importance of documentation, the approaches to documentation, as well as to all issues related to testing of instruments. It's important, I believe, for the industry as a whole to make an investment in standards and in tools. I don't think we should be off doing this on our own; we need to be doing this as an industry. I think that the burden of documentation and testing can be reduced through industry-wide standards, and I hope that we can begin the process of setting those standards as part of this workshop.

So [those are] my opening remarks, and Chet apologizes for not being able to be here, but there was the minor issue of the budget for the U.S. Census Bureau, so I guess we can forgive him. If there are any questions about the overall focus, we could entertain those, or I could move right into the talk to provide the setup for the problem. [*No questions were raised.*]

WHAT MAKES THE CAI TESTING AND DOCUMENTATION PROBLEMS SO HARD TO SOLVE?

Pat Doyle

DOYLE: So, why is it that we now have a documentation and testing problem that we find so hard to solve? Even though we've actually been in the area of automation of surveys since the mid-1980s, we didn't really face these problems until much more recently.

I'm sure most of you know by now what we mean by computer-assisted personal interviewing, but let's try to get everyone on the same basic page on what we're really talking about. Basically we have a set of predetermined questions. The phrasing and the sequence have already been set up, and they follow explicit rules that are conditioned on the circumstances of the respondent.

The question text and the rules are implemented in an interactive transaction-oriented computer system. And the system allows for prescribed flows as well as unprescribed flows, i.e., you can back up and go pretty much anywhere you want.

And that ability to back up and move forward and back up creates an infinite number of unique instruments for any given instrument that we can field. Obviously we don't implement any more than we have respondents to talk to, but we have the capacity to have huge numbers of different variations on the instrument. And the important thing is that all of these have to work perfectly, because we don't know until we get into the field which unique path has to be functioning for a given observation.

The Census Bureau is known for doing complex surveys. We have always done them. We did challenging surveys in paper; we've done challenging surveys in computer-assisted telephone interviewing (CATI). But, boy, now that we're in computer-assisted personal interviewing we *really* have some complex instruments, one of which is the Survey of Income and Program Participation (SIPP). [*For illustrative purposes, a one-page excerpt from a pen-and-paper version of the SIPP instrument is shown in Figure II-1.*]

SIPP is a longitudinal survey, which means that we go back to the same set of respondents repeatedly. It's nine rounds of interviewing for each sample that we have in the survey. We do three interviews a year; we go back every four months to the same people and also talk to whomever else they live with. We spread the interviewing out over time to equalize the field work load, so every month we're doing about one quarter of the sample.[3] We have an instrument that has about 13,000 different items in it that we've put out in the field, that we've automated. It's personal interviewing—carry the laptop to the house, or be at home with the laptop making telephone calls. The instrument is completely coded on the laptop, and there [are] some separate systems that control when the instrument comes in and when the case is set up and so forth.

The information that we collect is monthly, so for every month of the four-month reference period we collect information. We have data that we collect every wave, which is what we call "core information." We have that for income—well over 50 different income sources. Extensive demographic characteristics. A lot of work history and activities, so we

---

[3]The sample of households to which SIPP is administered is known as a "panel." To divide up the workload, the group is broken into four subsamples; one of these "rotation groups" or "rotations" is interviewed each month, hence each panel member is interviewed every four months. At that interview, the respondent is asked to report based on the preceding four months; this four-month window is the "reference period" for the survey. Each round of interviewing for a particular household is known as a "wave," so that the first time the interviewer contacts the household is called Wave 1, the second contact Wave 2, and so forth. For additional detail on the SIPP and its sample design, see Westat (2001) and National Research Council (1993).

Section 3 – AMOUNTS

Part A – GENERAL AMOUNTS (ISS Codes 1–56)

1. You said . . . received (was authorized to receive) *(Read name of income type)* **during the 4-month period.**

(Read "was authorized to receive" if asking about "Food Stamps" – code 27.)

Income code | Name of income type

3200

CHECK ITEM A1 — *Mark (X) income type code.*

3202
1 ☐ ISS Code 1 or 2 (SS or RR)
2 ☐ ISS Code 25 (WIC) – *SKIP to 13a, page 31*
3 ☐ ISS Code 27 (Food Stamps) – *SKIP to 11a, page 30*
4 ☐ ISS Codes 37, 50, 51, 52, 53, or 56 – *SKIP to Check Item A4*
5 ☐ Other ISS Codes – *SKIP to Check Item A4.1*

CHECK ITEM A2 — *Refer to cc item 27.*
Is . . . a designated parent or guardian of children under age 18?

3204
1 ☐ Yes
2 ☐ No – *SKIP to Check Item A3*

---

2. During this 4-month period, were any separate payments from (Social Security/ Railroad Retirement) received especially for . . .'s children?

3206
1 ☐ Yes
2 ☐ No – *SKIP to Check Item A3*

---

3. Did . . . also receive a separate payment for (himself/herself) during any of these months?

3208
1 ☐ Yes
2 ☐ No – *SKIP to 9a, page 30*

CHECK ITEM A3 — *Refer to cc item 26a.*
Is . . . married?

3210
1 ☐ Yes
2 ☐ No – *SKIP to Check Item A4.1*

---

4. Did . . . receive (Social Security/Railroad Retirement) jointly with . . .'s spouse?

3212
1 ☐ Yes
2 ☐ No – *SKIP to Check Item A4.1*

CHECK ITEM A4 — Has information about the amount received by . . . from the income source entered in item 1 already been recorded during an interview for . . .'s spouse?

3214
1 ☐ Yes – *SKIP to next ISS Code or Check Item P1, page 53*
2 ☐ No

CHECK ITEM A4.1 — *Refer to item 11b, page 5.*
Is this income source listed on the income roster?

3215
1 ☐ Yes – *ASK 5b*
2 ☐ No – *ASK 5a*

5a. In which month, during the 4-month reference period, did . . . begin to receive *(Read name of income type)*?

Mark "Yes" in item 5b for the first month received and mark "No" for the previous months. Then ask if it was received in each of the remaining months of the reference period and mark item 5b.

b. Did . . . receive any *(Read name of income type)* **in** *(Read each month)*?

NOTE – Social Security and SSI payments may be adjusted for inflation each January.

5c. Some persons receive more than one payment per month for certain income types.

▶ *For ISS codes 1 or 2 (SS or RR) read –*

How much did . . . receive in *(Read each month marked "Yes" in item 5b)*? **Please answer by giving the total amount each month AFTER any deductions such as Medicare premiums.**

▶ *For all other ISS codes read –*

How much did . . . receive in *(Read each month marked "Yes" in item 5b)*? **Please answer by giving the total amount each month BEFORE any deductions.**

(Last month)	3216 — 1 ☐ Yes / 2 ☐ No / x1 ☐ DK	3218 — $ ____ . 00 / x1 ☐ DK / x2 ☐ Ref.
(2 months ago)	3220 — 1 ☐ Yes / 2 ☐ No / x1 ☐ DK	3222 — $ ____ . 00 / x1 ☐ DK / x2 ☐ Ref.
(3 months ago)	3224 — 1 ☐ Yes / 2 ☐ No / x1 ☐ DK	3226 — $ ____ . 00 / x1 ☐ DK / x2 ☐ Ref.
(4 months ago)	3228 — 1 ☐ Yes / 2 ☐ No / x1 ☐ DK	3230 — $ ____ . 00 / x1 ☐ DK / x2 ☐ Ref.

FORM SIPP-13200 (11-12-92)

Figure II-1 One-page excerpt (out of 63) from the core questionnaire document, Wave 3 of the Survey of Income and Program Participation (SIPP), 1993 Panel.

can get a good measure of earnings. We also have questions that vary from one round of interviewing to the next, so in one interview we might ask the respondents about their asset holdings and in another we might ask them about their fertility or their marital history. Those are called "topical modules," so they vary from one round of interviewing to the next.

One of the things that helps complicate our instrument development is that when we're doing a longitudinal survey we don't like to look dumb. We like to have our [field representatives (FRs)] look like they remember they were here four months ago and be able to say, "Oh, the last time you told us you were doing *x*; are you still doing *x*?" That's called "dependent interviewing." It's used in some cases to reduce some spurious error that might occur because a respondent might reply slightly differently now than they did last time—a lot of characteristics like industry or occupation where we can get a slightly different answer if they use a slightly different set of words even though nothing else changed. It's also used to control for—smooth out—what we call the "seam bias." For those of you who don't know: remember I said we have four months in each round of interviewing, so you can get a transition on and off a program from month one to two, and then two to three and three to four. And then you have another round of interviewing, so you get month five from the next wave. Well, so now you have transitions one to two, two to three, three to four, *four to five*—you'll notice how I said, "four to five"— five to six, six to seven. What happens is most of the transitions tend to be reported at the juncture between the rounds of interviewing, instead of evenly pretty much throughout the period—that's what's called the "seam bias."

Another survey where we're in the process of instituting an extremely complex instrument is the Consumer Expenditure Survey (CES). This survey is used as a component of determining the Consumer Price Index (CPI). We have four quarterly visits, plus a bounding interview, to households. Now these are addresses that are followed, so they go back to the same place every time—they don't always talk to the same set of people. This particular instrument only has about 200,000 separate items that it has in its store of things to ask. It takes an hour and a half to two hours to administer. If you can imagine you have all your expenditures, we're trying to get a complete record of all of those items and how much you spent on them. It basically covers all sales and excise taxes for all items purchased for basically the unit we're interviewing and for others, and it's all classified by type of expenditure. So you can see that it's an *extremely* complicated instrument.

For those of you who are not familiar with the way we do these surveys, I'd like to give you a couple of illustrations as to why we have a complex problem. Let's consider a question about the new [State] Child's Health Insurance Program (SCHIP).[4] This was instituted a few years ago in response to the concern about lower-income children not having adequate health insurance coverage. So, you might think to yourself, "Let me figure out who's covered by this program." So you have a simple question—"Is ⋯ covered under the new child's health insurance program, in a certain period of time?" Simple, basic question. You might think to yourself, "So what's so hard about putting that into a computer and bringing it up on the screen?" Well, if that's all we did, that would be fairly simple. And when we were in paper our questions usually weren't a whole lot more complicated than that. We left it to the field representative to do a lot of the tailoring of the questions to the circumstances of the respondent. So the FR would know to fill in "⋯" with either "you" or the person's name—they'd be trained to do this, they'd have the responsibility. The respondent would be trained in how to fill in the reference period, so they would know that if they were in SIPP Rotation x, that they would fill in the four months that were appropriate for that interviewer. So the instrument just had it loosey-goosey there, so to speak.

Well, now that we've got it automated, we don't leave it like that anymore. We now take the opportunity to make this question look different for virtually every respondent. So, one thing we do is to first figure out whether we're about to talk to a small child—well, we don't talk to small children directly, we talk to their parents. So we have to have a version of this question for the parent, and that question might be, "Are any of your children covered under this program?" Or it might be a separate question for every child, where we maintain a roster of children and the computer will cycle through the question, explicitly naming each child. We might condition the phrasing of this question on whether they've had any other kind of insurance—"you've just told us that you have insurance type x; now, do you also have this type of insurance?" As I mentioned earlier, we want to look smart—we want our FRs to look smart, to acknowledge that [the respondent] told us they were covered by this last time. So we would phrase the question, "Last time you told us Sally was covered under CHIP. Is Sally still covered under CHIP?" As I noted, we want to get the reference period right, and if it floats like it does for some surveys the computer is going to predetermine what it is.

---

[4]The background and analytic needs of the SCHIP program were recently the focus of a Committee on National Statistics panel (National Research Council, 2002).

Another thing that we like to do which we couldn't do under paper is [that]—for these programs that have different names depending on the locality, which is true for most of the welfare programs now—we want to pre-load all those names, and check the address of the respondent, and instead of saying "CHIP" we might use the explicit name for the program as it exists at that particular location. We might also want to condition this question on their income characteristics, because CHIP is also a program for low-income children. So maybe we wouldn't want to ask the question of really rich households, or maybe we want to ask it a different way. We also have situations where we accept a proxy response, so we would want to phrase our question a little differently depending on whether or not we're talking to the respondent or talking to the proxy. And then of course there's always the need to have some facility for the FR to answer a respondent's question: "[*What*] is this program you're asking about?" So we need to have that sitting around.

So there you are. Now we've taken one simple little one-liner question and created probably three pages of computer code of some kind or another to carry out all the variations on a theme that one could *possibly* consider one *might* want to have in the field. The interesting thing is that what you're trying to do is condition the question on the characteristics of the respondent before you know what the characteristics of the respondent are. So you're back here imagining all possible things that could happen instead of being able to just precisely target it to what you're going to get in the field—you have to anticipate it all.

Let's look at one more example. So we have a fairly simple question— we want to know what their income was from a particular source; in this case it's an asset-type source. Again, in the original paper version of this, we have a fairly simple question; we have the ellipses where we would fill in the respondent's name, or "you", or "he", depending on the circumstances—again, up to the FR to decide, a lot of freedom there on their part. We'd probably in this case have this set up with a little list of income sources that they've compiled—the respondent has gone through and said four or five different things, so now they've got a little list. And they'll cycle through this question—and they have instructions to tell them how to cycle through the question—going through each of the four or five income sources that they've got. Now we go to automated— we no longer have the simple one-liner. We would vary the question depending on what type of account they have. We would—particularly in the case of SIPP—when we ask for amounts, we ask the question differently depending on whether or not it's reported jointly. And, in the newer version, dependent on who it's jointly with. So, if you have account joint with your spouse we'll ask the amount question one way; if you have an account jointly with a minor child we'll ask the question a second way; if you have an account jointly with somebody else we'll ask it a third way; and if you have a combination of these, well, Lord knows what you're going to get. And, of course, if you just have a single-held account we would just ask you what the amount is.

Now we also use dependent interviewing in these circumstances. In this case, what we're looking at are situations where we don't initially report back what they told us last time—we wait and let them tell us something. If they don't know an answer, if they don't remember, then we do a check to ask them a slightly different version of the question: "Was it bigger than x?" Or, "was it between x and y?" And if they still don't know then we'll say, "The last time we were here you told it was about $100. Is that right? Is that about right?" So that will trigger a memory, hopefully, that they can use.

So there we have our infinite variations on this particular theme. And this is—remember—one of 13,000 different items in the SIPP survey.

So basically what we're doing is trying to optimize the question. We want it to be clear to the respondent what it is we're trying to ask. For example, if you're talking about income from a job you probably want to call it "earnings" instead of "income," just because that's what we're used to using in that context. We want to find the easiest way for the respondent to answer. One of the things we're experimenting with in SIPP now is to ask the respondent whether it would be easier for them to report monthly, weekly, biweekly, annually—whatever way they like, and then depending on which of those they choose they get a slightly different set of questions in order to answer the amount. We'd like to create a friendly rapport between the field representative—which is an interviewer, by the way; this is Census Bureau-speak for an interviewer, my apologies—we want that interaction to be friendly, so we don't want to force the respondent to answer a question in a way he doesn't want to. In this case, we wouldn't force them to report monthly if they really didn't want to. If they want to report annually, well, we can divide by 12—no problem, we're in a computing environment, right? The computer can do it. And we also want to give them lots of opportunities to report things in alternative ways.

OK, we're still not done with the level of complexity that we can introduce into this instrument. We now have all different types of question structures, answer structures. We can have a simple "yes"/"no", and with a simple "yes"/"no" we can allow them to have a "don't know/refuse" or not to have a "don't know/refuse". We can have a "don't know/refuse" be visible on the screen or invisible on the screen. Lots of options, just for a "yes"/"no" answer. We can have a single question that has several different answers—if we ask them how long they've been involved in an activity, we might let them fill in days, months, or years. So we would have three different answer opportunities. We can put several questions on a screen or one question on a screen. We can have a different style of answer category if we expect the response to be continuous—like, "give me your income"—versus discrete—"do you have white hair or purple hair?" We also have to deal with a numeric/alphanumeric mix—whatever we do, we allow "don't know/refuse", and that comes in as "D" and "R" sitting in a field with amounts like "$100" and "$150". And of course we can never anticipate every possible category, so we always have an "Other (specify)" with our other categorical responses.

BANKS: I don't mean to interrupt, but "don't know/refuse" get conflated—they're not separate responses?

DOYLE: That's correct. You can design it pretty much any way you want to, but it's sitting...

BANKS: So implicitly someone who just says they "don't know" is the same as someone who...

DOYLE: Oh, no, no—they're separate. I misunderstood your question. You get a "D" or an "R"—my point [is that] it's the mix of alphanumeric and numeric that causes some data processing procedures to get a little unhappy.

Sometimes we want to change our questions slightly depending on the way people answer, so we have these little pop-up things—I don't know what you call these things in a cartoon, but they just pop up. And I mentioned the "other (specify)" and "don't know/refuse" options.

OK—I'm still not done yet. Let's talk about what causes the infinite variety of these questionnaires. At any point in an instrument, the Census Bureau—and I know not everyone implements it quite this way—but at any point in an instrument our field representatives can back up and change an answer that was previously made. There are no blocks in there; there is no place that stops them. I know some instruments will limit how far back you can go, but we can go back. We can go back to the prior question, change the answer, and come back to the current one. We can go back five or six different questions, change that answer, come back to the current one. We can go back 10 or 20 questions, change that, go down a slightly different path, and then come back to where we started. Or we can go way back, down a different path, and never get back to where we started. So, we have to allow for that kind of unanticipated flow through an instrument.

PARTICIPANT: When you go back, is it just as though you had never gone down that path? So if you go back to a question that has a branching characteristic, and you've made response to that, do you in effect discard all the former material that had been collected so [that] it's as though you never went down that path? Or do you [use] some of that...

DOYLE: It partly depends on the implementation, but it also depends on the system. CASES, anyway, doesn't erase it automatically;[5] theoretically, it's eventually supposed to erase it if you don't go back down that path again. But it's still sitting there waiting for you because it doesn't know quite yet whether you want to change. So if you don't change the path the answers are still there; you don't have to re-enter them to keep on the same path. But you shouldn't have to go back and discard them when you change it, but we've had some issues with the CASES software. Now I don't have experience with Blaise.[6] Ed, you're here; can you talk a little bit about your experience with Blaise and the backing up. Does it reset the on-path/off-path sort of stuff?

DYER: How much time do you have? [*laughter*] We've run into a lot of problems with setting outcome codes, for example, just recently where if you set it as you go through the normal path it's there for you but if you back up it doesn't erase them unless you tell it to do so. You can arrange it to have those reset; you can initialize those variables. Now there's an issue about taking it all around. [But more than that] I would have to hear from the experts.

PIERZCHALA: [In Blaise,] if you back up and there are off-route data when you close the interview, the off-route data are wiped out at the time you close the interview. But they're available while you're still in that session; they're there, so if you come back to that path and discover that you should have been there all along they are still available and you don't have to re-enter them.

PARTICIPANT: That's probably the optimal solution.

DYER: ...unless you happen to design that thing with auxiliary variables and let that off-path get reset, so it's really based on what your intent is. You can do any number of things.

PIERZCHALA: Well, that's what I'm saying—you can program any number of things, and you have to be careful what you're going to program in the first place. And my experience is that a lot of people who are complicated in what they program get mixed up backwards.

DOYLE: Yes, Tom?

---

[5]CASES—the Computer-Assisted Survey Extraction System—is one of the major software environments currently in use for developing computer-assisted interviewing questionnaires. Developed and maintained by the Computer-Assisted Survey Methods program at the University of California at Berkeley, CASES is a DOS- or Unix-based package, with a Windows version still in development.

[6]Blaise is a Windows-based system for authoring and managing computer-assisted survey instruments. Blaise and CASES represent perhaps the two dominant survey software packages used by survey organizations (particularly among U.S. federal statistical agencies) at the time of the workshop. Developed by Statistics Netherlands, Blaise is distributed and marketed within the United States by Westat.

PIAZZA: Just to clarify what CASES would do in that situation: basically, it leaves the data there until you go through a cleaning process, and then it will wipe it out. But often [the] Census Bureau doesn't clean their instruments anyway, so sometimes you have the strangest things because you don't go through that step.

DOYLE: OK, so does everyone believe that this backing-up feature is complicating our instruments?

The other thing we need to be able to do is to successfully—abruptly—end an interview in the middle of talking to a respondent. I mean, things happen, and they stop talking, and you say, "OK, fine, thank you," and go on to the next person. [This] needs to be able to happen without destroying the instrument. We also need to be able to do this for a whole group of respondents in our case. And once we've done that we have to start all over again in the middle. Yes?

ROBINSON: And when you abruptly end, what happens to the data that was entered so far?

DOYLE: I need an expert to answer that question. Adrienne?

ONETO: The data that was entered so far is saved.

But I just want to make a comment, though. A lot of what you've described so far really is our choice of how we let the automation. I mean, we could have still continued to use our interviewers to determine universes and to determine skip patterns, and to decide what is the most appropriate fill for a question. I think there was this insistence among all of us that once we went to automation we were going to use automation to replace this very complicated cognitive process that all interviewers go through when they conduct the interview. We always had multiple question types, we always had very difficult question puts—I mean, a lot of what you described always existed with our paper instruments. It's how we chose to implement the automation, and I think that's important. We could go back to the sort of paper approach and let the field reps determine the universe and the skip and the flow and the . . .

DOYLE: Except for the cognitive people of the world, who have now got a gold mine for ways that they can predetermine how to appropriately phrase the question so that the FR isn't using a little bit *too* much imagination in determining how to phrase it. So it's good and bad, but you're right. A lot of this is not because we automated; it's because we chose to do automation in a certain way. And we're not alone. Most people are choosing to do automation in this way.

PIAZZA: Let me make a comment on that. Although it's true that we could leave a lot of fills to the interviewers, when you're talking about flow it's a little trickier. Remember, the interviewer can't just flip the pages of the questionnaire.

ONETO: Right.

PIAZZA: So the flow really has to be built into the instrument. And that, in a sense, was part of the attraction of the computer-assisted interviewing in the first place—that somehow we would eliminate these errors in flow.

ONETO: Yes, that's true. Except that we could still do a traditional check item, where they're asked a lot of things and that decides what happens next. But we don't do that. We collect information that's already been provided and—behind the scenes—we do the check item and the flow and the skip. That's what I meant.

PARTICIPANT: Do you ever get any pushback from some of the FRs, especially perhaps those who are more experienced or maybe those with no experience, that they feel they can determine the flow or some of the questions better than the computer can, or that they lack the flexibility that they would get with a paper instrument where they could just flip the pages? Does that ever happen?

ONETO: I think it's quite the opposite. Interestingly enough, at the Census Bureau, at the nascent states of automation they resisted it. But now the interviewers who are trained with automation do *not* want to go back to paper, they don't want the burden of skips and universes and everything else. They like the fact that the automated instrument does it for you.

DOYLE: Yes, basically, that's been our experience. Now, they do have different attitudes about what works best in the field than we do at headquarters. So we use them in the instrument development process— it's not quite a formal usability test, but they can sort of tell whether this thing is flowing well or not. And we'll go back and modify the behind-the-scenes to allow it to flow better and be more comfortable for them. But they're not asking us to do that control themselves.

COHEN: Could you define "universe", "fill", and "skip", for those of us . . .

DOYLE: Sure. A "fill"—if you remember back to the question we had before, "What was ···'s interest income in reference period, from [some] source?" "···" is a fill. In other words, there would be an instruction behind the scenes to fill in either "you" or the person's name or "he" or "she", depending on what the instrument knows about the respondent at that time. That would be a fill; it's like a piece of code that basically does a calculation based on a characteristic and plugs in a name or something. Similarly, reference period could be handled as a fill, so that instead of seeing a parenthetical remark describing a generic reference period, when this question goes up on the screen it would say "April, May, June, July," or it would say, "since April 1". In other words the computer has changed the appearance or the wording on-screen so that it meets the need of this particular case. That's the fill.

The skip is . . . Let's say that just before this interest income amount question we had a question that said, "Do you have an interest-bearing account?" And if they say no we might have some logic that says, "OK, don't ask the next amount question—skip it and go on to the next." That's a skip.

Then the amount question itself now has a universe of people who have said, "Yes, I have an account," so therefore they're going to get to answer the amount question. So that's basically the universe; the skip and the universe go together. The skip patterns tell you how to go forward through the instrument, and the universe tells you who was allowed to answer this question.

OK. So, we're still not done yet in terms of describing how we can make this thing complicated. We don't always literally record the output the way it was input. So when the FR types in a number we may or may not store that number in our database. Now, sometimes we don't store it because there's a bug in the instrument. Sometimes we don't store it because we didn't *intend* to store it; we might store a recode of it, or we might take two or three questions that are really getting at the same thing and only store one output. We have the freedom to do this if we want. As we said before, we can have very mixed-format output fields that kind of drive computer programs nuts. But we have to make sure that they're correctly placed on the files. We also have inputs to the instrument. Whenever we go out we don't just start with a blank slate; we have a little file coming into our instrument that's telling us where the address is. If this is the second time we've interviewed them, it's feeding back data from the prior interview. So we've got this input file we've got to worry about, getting it right and getting the interaction between it and the instrument to work right.

We also have the infamous systems. The instrument doesn't operate in a vacuum; it operates within the context of some other systems that keep track of what cases should be interviewed, which cases are assigned to which field representative and where the cases are; whether they've been spawned or not and, if they are, which interviewer gets sent off to them. And then of course there's the all-important system that captures the data after the interview's complete and sends it back to some database somewhere, housed in somebody's computer somewhere. It also is the thing that keeps track of when you've abruptly stopped an interview and have to restart it; that's what's controlling that activity.

On top of all that we do what automation was sort of built up to do— we do online edits. In other words, while we're asking these questions, if the respondent gives us an answer that we don't think is right, we can check it and we can ask them again. "Are you *sure* you meant to tell me that?" Now, we don't do that very often because it doesn't exactly go toward friendly interactions with the respondent. But we do it a lot when we're looking for FR keying errors, for example. If we expected a number in the range 1 to 99, and they key in something like 10,000, we might come back with an FR message screen asking whether they can go back and doublecheck that, whether they keyed that in correctly. So we do a lot of that sort of range-checking on the dates; if they say they got married before they were born, we might want to clarify which of those dates was the correct one.

So why is this now a problem when it wasn't a problem before? When we were doing instruments in paper we were using a medium that constrained our options to those that humans are capable of understanding, capable of documenting, and capable of testing. When we initially did computer-assisted interviewing we were doing it in a telephone environment, often in multi-mode surveys. And the computer-assisted telephone interview (CATI) questionnaire would often track the paper questionnaire. So we were implicitly using the "KISS" [*"Keep It Simple, Stupid"*] method of instrument design. We still had a lot of responsibility out there for the field rep to do the fills; the flows were not very complicated. We could easily tell what all the possible paths were. We could print out the instrument. We had documentation, and so forth. But when we got to computer-assisted personal interviewing and we weren't doing multi-mode surveys, boy, we went to town!

And, as Adrienne [Oneto] said, it's not that automation makes it worse; it's that we are allowed to make it worse because we are automated and have no constraints.

So, I believe the solution is that the industry needs to restrain itself. But, also being an instrument designer, I know that it's absolutely impossible. When I sit down to design *my* set of questions that I want to go into the field with, I'm as bad as anybody else. "No, I'd like it this way if it's this set of circumstances, and this way if it's that set of circumstances." And I can't restrain myself any better than anybody else can.

I was going to talk a little about the documentation problem and then about the testing problem separately. Yes?

BANKS: It seems that there are sort of two extreme positions that one could take. One is that you you want to presume, to grow, to plan [for] all possible contingencies and plan for that whole tree of possible questions. And the other approach is to basically allow the FR to have a conversation with the respondent and then to fill out the form in an intelligent fashion based on what the respondent said. Is there any experiment that has been done that compares the yield or the accuracy of results under those two kinds of positions?

DOYLE: Would you like to answer that question?

PARTICIPANT: Well, this builds on work that I've done on the conversational literature. Not exactly conversations, but there have been experiments where interviewers can override, where they don't *exactly* follow the skip patterns—it doesn't allow full improvisation, but it does relax the script. And in some cases that does dramatically improve accuracy. Most of that comes from clarifying concepts, from clarifying what's being asked in various parts of the script. That's not exactly what you ask because there's not much relaxation of the skip patterns.

GROVES: On the other hand, there's a paradigm in measurement that asserts that the sensible thing is to standardize—rightly or wrongly—because when you get the data back then the completion of the stimulus accrues to the measurement.

DOYLE: Mark, you had a point?

PIERZCHALA: Well, in all the papers I've read about accuracy, the one place where computer-assisted interviewing has really improved things is keeping people on the correct path. You're not asking questions that you should not ask; you are asking the questions you should ask. On paper we have both those sorts of errors. You get data on paths they shouldn't have gone down and no data on paths where they should. So we want to keep the skips in there. But there's this whole different aspect—do you really need to carry the fills out to the nth degree? And can't the field representative—with their native intelligence—provide the fills themselves?

PARTICIPANT: The issue of restraint is an interesting one, and I'm just asking whether it comes back to negotiating a dialogue between the people responsible for system engineering and testing. It involves saying, "Look, we want certain standards; this is how the surveys are all going to be organized." And then on the other hand you have the people who do the survey design and—with their set of issues in mind—push the envelope against that. Is that the way it is, and if so how do you negotiate that? Is that negotiated in advance, or each time it comes up?

DOYLE: It's unique to the circumstance. Lots of times either the budget or the timeline will drive what you can do. So, if you've got your survey designers *really* making this thing super-precise, and it's getting really complicated, and then all of a sudden the office sees it and says, "Oh, my goodness, there's no way I'm going to be able to author this in the two weeks I've got left," then we go back and say, "No, no, no, no, no—we can't do this." So there's some back and forth, and there's some middle-layer people who do the negotiation between the designers and the authors. Lots of people design these staffs differently, but if you think of three different functions there, that's typically what goes on.

But there are other ways we can restrain ourselves without getting away from the precision, and that's just to simply have standards for straightforward things. Like, for instance, "his" and "her" fills—there's *no* need for anybody to ever have to program a "his" and "her" fill, after it's been programmed one time. There's just no need to do that. If everybody agreed that this is how we do a "his" and "her" fill, then there's a little function out there and you just have to plop it in. So that's a level of restraint that's different from how complex can we make the question. [Would] I give up my freedom to do a "his" and "her" fill the way *I* want to do it in favor of the way the majority of people want to do it? Which is a different kind of restraint, you see.

PARTICIPANT: This may be something you glossed over in describing the complexity in case management: What kind of flexibility is given to the interviewers in terms of the outcome of the case? What happens with it, and with the Census Bureau's philosophy of all behind-the-scenes, all automated? Do other organizations think of the interview the same, what the outcome is? At the end when you're done? That carries a lot into the development process and the testing.

DOYLE: The whole systems side is something I haven't dwelled on, and I sort of view the setting of that—maybe I'm wrong, but I kind of put that final outcome code setting more on the systems end.

For those of you who don't know: when the instrument is on the computer, there are interim stages—for instance, you've opened the instrument, you've tried to conduct the interview, you weren't successful. You make some sort of interim record that you made this call. Maybe you'll try again later. You come back later; maybe you got half the interview done. You come back eventually, you get the whole interview. And you make a record. And, at the end, that record of what finally happened is the final outcome code. And that's what Bill's talking about. And there [are] some choices that can be made, as to whether you try to keep track of every one of those stages or whether to keep track of the final stage. And the Census Bureau has tended to only keep track of the final stage. But we've now decided that that's not in our interest; we want to know more about what's going on [in the] attempt to get that interview, so we want to go back—and increase our complexity—to retain all the various different interim codes that came along during the process. And some of the information comes from having opened the instrument; some of it comes from case management, where you don't even open the instrument. You drive by and discover that you can't get into the house; you don't open the instrument, and you have a record in your case management system that says that I tried to interview but I couldn't.

PARTICIPANT: Just another comment, with regard to [computer-assisted telephone instruments]—they tended to be simple, but then again the automated call scheduling system was extremely complicated.

So you have a very difficult testing problem, to get those to work correctly. And if you're going to change your algorithm ...

DOYLE: And is that still true?

PARTICIPANT: Yes.

DOYLE: So, you can tell my bias is more towards the guts of this thing. And less toward the outer pieces that make this operation work—both of which are important. Mark?

PIERZCHALA: But getting back to the guts of the thing, there's one thing you haven't mentioned with respect to this, and that's software in two or three languages.

DOYLE: That's true, that's true.

PIERZCHALA: The pattern of the way you fill with French and English are different. The grammatical structures of the sentences are done differently. And there [are] instruments out there in the world where you don't know what languages are going to be done.

DOYLE: You're absolutely right. I completely omitted that. So add yet another dimension to this entire problem.

I don't think we're ever going to take any dimensions away; we're just going to add them.

So, how are we doing on time?

CORK: It's about five minutes until ten, so you've got about thirty-five minutes left.

DOYLE: So I want to do documentation, and then I want to do testing. So I'll do half-and-half.

So why do we have a documentation problem? First of all, when I talk about documentation now, I'm just talking about documenting the questions that were asked in the field. I'm not talking about documenting files; I'm not talking about study-level quality profiles. I'm just talking about documenting the questions used out in the field.

The main reason we have a problem is that we no longer have the free good of a self-documenting instrument. And we have limited resources to manually develop a substitute for that missing instrument. As you might expect, an organization whose primary purpose is collecting data is going to put [higher] priority on collecting more data and collecting it better than they are on documenting what they collected.

We have some attempts now at automating tools to do documentation. They're still fairly new; I think we need to continue pushing in that area. Even when we do succeed in documenting an instrument, the thing is *so* big and *so* complicated that you can't get your brain around it. One of the things that we've been most weak on in our documentation attempts so far is to provide the context for each question.

POORE: Why do you want documentation? [*laughter*] If you're not going to be able to get your brain around it, and if it's so hard to do, why do you even want to do it?

SCHECHTER: That's why we like it. [*laughter*]

DOYLE: Yes, that's right; partly it's for Susan [Schechter and the Office of Management and Budget (OMB) review]. In order to get our questionnaires approved to go in the field, the people who approve them have to be able to know what's in them. So that's one reason.

POORE: But if they can't get their brain around it, then they can't go out in the field?

DOYLE: No, they do, it's just ... you can't do it in its entirety. What you do is to break it into pieces. And so this individual will sort of know this piece at some level, and this individual will know that piece at some level. And what you can't do is to get the whole thing and—more importantly—you can't get all the nuances. Now, when Susan approves an instrument, she hasn't seen all the nuances. She doesn't see all the fills. She'd never get anything cleared. We'd never be able to do another survey, if she had to sit and look at every possible variation on a theme.

POORE: In that circumstance, how are you able to ascertain that the document does the job it's intended to do?

DOYLE: It's difficult. One of the reasons we have a couple of attempts to do automated instrument documentation is so that we know what we're getting as a documentation of the instrument as opposed to what we intended to collect. We have other tools that can document what we intend to collect.

POORE: Is that a subjective judgment, or is it ...

DOYLE: What do you mean? Is it my subjective judgment that ...

POORE: That someone decides that document does what it's intended to do?

DOYLE: No, usually it's, "I'm out of money; it's the best I can do." Seriously. We've run out of money, we've run out of time. This is what we've got; sorry, that's it. It's the best we can do. We know it has problems; maybe if we had more time and money we could fix it.

GROVES: This seems to be a fair question, and it could be revisited. I believe you've implicitly put the word "paper" in front of "documentation". If the intent is to get to the other person information about what the instrument is, does it necessarily have to be on paper?

DOYLE: No, nor do I think it has to be on paper. I didn't mean to give that impression. In fact, I think it's impossible to do on paper. What we need is paper providing some amount of information but we can't rely on paper exclusively because it's just too complicated to represent that way.

BANKS: I'm really sympathetic to the point of view that was just raised. It seems to me that if I were trying to get approval for a new instrument of this kind, I might want to know whether 80 percent of the respondents fall into a set of about 10 pathways that are truly understood and the remaining 20 percent are in some incredibly complicated, chaotic situation that's going to be so individually variable that it's going to be very hard to map. I guess I would feel more comfortable if I knew that some large proportion of the respondents were simple and well-captured and the real complexity only applied to a small percentage.

DOYLE: That's generally not the criterion that's used to clear an instrument. The criterion is largely burden—how many questions are we administering these people, how many minutes are we going to be in their home, bothering them to collect this data? And is any of this so redundant with some other collection that we couldn't save some burden by . . .

BANKS: But I think this suggestion goes directly to that, because if the 10 pathways that 80 percent of people employ in fact would cause little burden, then that's helpful to know.

DOYLE: But typically it goes the other way; the most complex paths give you the biggest break on burden, because you've tailored it so specifically that you're only really asking the question when you need to.

MARKOSIAN: So is there a concept of an acceptance test for an instrument, and in particular is there a component of the acceptance test that sets general criteria that need to be passed? Is it institutionalized? Is that kind of written down anywhere?

DOYLE: I would say no.

MARKOSIAN: For example, complexity in terms of actual paths through the instrument.

DOYLE: No, no. First of all, we haven't been able to measure that.

MARKOSIAN: Well, I don't mean just that one issue. Surely, before you sign off on an instrument you have some criteria in mind about what that instrument needs to do.

PARTICIPANT: I think that acceptance testing is one of the most difficult challenges of automated surveys in general. It's very iterative, it always takes a lot more time than anybody ever thinks it should. And I think you're going to go into some detail on testing. I mean, it's multistage—you do a first clean-up, you do a cosmetic. Then you get into the logic and you just go deeper and deeper into the instrument to make sure it's field-worthy.

DOYLE: And there's different types of acceptance. There's acceptance in terms of, "OK, I'm asking for the information I want to get." That's what the analyst is looking for—"Does it ask for what I want?" Then there's acceptance from the cognitive people—"Is this question being phrased in a way that the respondent can understand it?" Then there's acceptance from the operational side—"Is this thing going to break when my FRs are in the field, or not?" "Is it going to go down the right path, or blow up if I back up too far?" There's acceptance from the clearance process—"What is the burden, and how long is it going to take?" So there's a whole different set of criteria that's used to judge whether an instrument is ready to go and doing what it's supposed to do. And then there's also—"Can we afford it? Is this thing going to be able to be done within the amount of money we have for the field budget?"

McCABE: "Complexity" as a dimension, or perhaps as a criterion—is that just a general concept or is there a specific mathematical definition?

DOYLE: No—it's just a sort of vague nation that this is complicated, and this is simple; you can recognize those two but not really tell what's between them. Bill, there's a person behind you whose name I don't know . . .

PARTICIPANT: The slide [suggests] that the documentation occurs after the instrument has been completed, but in fact it's more of a specification problem. Because if you were able to write proper specifications during the development and before the development then your documentation would be less of a problem; it would come from the specs.

DOYLE: That's right, and in recent times we've moved toward using database management systems to develop the specifications, to automatically feed the software that runs the instruments. What we're finding is that it's very difficult for that to flow naturally. At that stage, people will get on the phone and say, "Oh, this isn't quite right, can you fix this?" And all of a sudden we've got a disconnect. We've got our specs maybe a week, two weeks, a month behind where we are developing the instrument. Maybe we have the staff to go back and put them together, but probably not; they're probably testing this thing, working really hard to get it into the field, and once it's in the field they're off to the next survey. That's where the problem is. But you're right; we can do better in terms of having a good set of specifications, and we are. Every time we do a new survey we're better than we were the last time we did it. But I'm pretty convinced we're never going to be able to perfectly document an instrument from a spec, unless we change our behavior.

That was the "continuous refinements are hard to track" part, and that's what causes the instrument and the spec to get out of synch. The other thing that could make our problem a lot easier is back to the old "his"/"her" fill—if we had industry-wide standards, and everybody did some of these things exactly the same way, we wouldn't have to document them every time we did them differently or had the fifteenth variation on the same fill. So if we had more conventions and standards

I think this problem would become more solvable.

Well, what are we doing now? Everybody produces basically what we call "screen shots"—a list of questions that are asked in the survey. Typically it's stylized in the sense that it doesn't tell you every possible fill. It tends to be written in a way that you can read it, rather than `TEMP1`, `TEMP2`, `TEMP3`, 5, 10—that's the way some of these things are coded. But it doesn't typically give you the flow of the instrument. Sometimes we do that, but a lot of times we don't. Flowcharts for instruments are pretty rare because we haven't yet really been able to generate them automatically; I mean, TADEQ has some approach to this that's under development.[7] We've had manual ones done but they're very resource-intensive to produce.

We have—as I mentioned—TADEQ; we also have IDOC.[8] Both of these are attempts to automate the instrument documentation process—to read the instrument software and produce information about that instrument. They take very different approaches. The IDOC notion was to basically capture all the little pieces of the instrument and put them into HTML documents, put it up in Netscape and let you use all the browsers and what-not to go through and follow the instrument. And if you wanted to know what a fill was you could go click on it and go back and find what it was. And if you wanted to find out what a flow was ... TADEQ is more focused on analyzing an instrument—telling you how many paths you have, telling you that a path is successful, and so forth. They're both really good starts at this problem, but we're not quite there yet.

I just mentioned the instrument development tools—we use these database management systems to do our spec-writing in the hopes that that will automate the instrument production and that we'll have better specs in the end. Yes?

COHEN: Could you use either IDOC or TADEQ to estimate the respondent burden?

DOYLE: No. No.

COHEN: So the only way to do that is to go out in the field and give it to 200 people and see how long ...

DOYLE: Well, and you can see how long it takes a couple of different ways. Now, with automation, we can put in what are called "timers"—you can literally see when the instrument is turned on and turned off and how long the time is. But that's a little misleading because the FR

---

[7]TADEQ stands for "Tool for the Analysis and Documentation of Electronic Questionnaires," and is discussed in greater detail in Jelke Bethlehem's presentation later in the proceedings.

[8]IDOC stands for "Instrument Document," and is discussed in greater detail in Tom Piazza's presentation later in the proceedings.

may not bother to turn it off when they leave. I mean, we have some timers of 36 hours that are clearly outliers.

MANNERS: I just wanted to say that one of the user requests that will be built into TADEQ is to allow the researcher to put in some times or weights to be built into the questions, so that those can be added up along a path.

DOYLE: Oh, OK.

At the Census Bureau we've produced some documentation guidelines for the demographic surveys. Basically, it's the mantra of "document early, update often." This is the behavior modification that I'd like to instill in *all* people who work on surveys. It affects all areas from analysts to methodologists, operational folks to authors, everybody—we need to get on this bandwagon or I don't think we can solve the problem.

We have some standards; we have them internally at the Census Bureau for the actual implementation of our instruments. We also have, in the documentation arena, the Data Documentation Initiative, which is producing standards for data documentation which includes standards for formatting instrument documentation.[9] That's going to help us a lot once we move toward really getting our instruments documented, having them in a format that can be shared in a machine-readable capacity across all the archives, all the data libraries, all the users, and so forth.

We've had three different attempts to produce recommendations for documentation. In the mid-1990s, the Association of Public Data Users (APDU) developed recommendations for the documentation of instruments. Then we had the TADEQ User Group in Europe produce some recommendations. Then there was a meeting at the FedCASIC conference—not this past year, but the one before that—where we talked again.[10] This time we brought in not only the experts in producing it but also a number of users; we brought in representation from OMB who does the clearance, our agencies who sponsor our surveys. In your booklet, I think it's page 24—there's Table 3 of an article in *Of Significance* . . . .[11] That article summarizes the recommendations from all three of these groups, and Table 3 has kind of got the snapshot picture of all those recommendations. I won't go through them now because I want to

---

[9]The Data Documentation Initiative is intended to foster an international standard for metadata (essentially, data about data) on datasets in the social and behavioral sciences. Additional information on the initiative can be found at http://www.icpsr.umich.edu/DDI.

[10]CASIC stands for "Computer-Assisted Survey Information Collection," and is a general term used to describe the community of organizations and agencies who conduct survey research using computerized instruments. An annual conference of the federal agencies involved in this work—typically hosted by the Bureau of Labor Statistics (BLS)—is known as the FedCASIC conference series.

[11]*Of Significance* . . . is the journal of the Association of Public Data Users. Specifically, the table Doyle refers to is in Doyle (2001).

move onto the testing thing. But it's pretty illustrative. It basically talks about these issues of whether we ought to have standards for "his"/"her" fills, the more straightforward fills. It talks about all the components of documenting an item.

PARTICIPANT: It's actually on page 115 of the book.

DOYLE: Oh, OK—page 115 of the book, page 23 of the actual paper.

The testing problem ... We need to test all of the systems—as Bill reminded me, they are quite complex, and I keep forgetting about them because they're not quite part of what I do. *Which is another part of the problem*, by the way. We have all these players in the process and none of them can quite understand why the other players need so [many] resources to do their job. "But *I* know why I need resources to do *my* part!"—so therefore I should have all the resources, and then we'll squeeze the authoring down a little bit, and—oh, well—we'll squeeze the testing down a little bit so we can do some of this other stuff.

And every one of us has the same problem; we don't do the other job, so we don't understand the other job well enough to appreciate their problem, their needs, or what I can do to make their life easier.

We need to test the input—the stuff that's feeding into the instrument to drive it. We need to test the instrument output—that was something that, in the early days of automation, people just didn't sort of think about. They just figured that if I'm typing it in, it'll be in there. And then sometime later they'd get a data file back and go, "Whoops—I missed a few variables here and there." So we learned our lesson the hard way.

We need to test the flow—all those paths, all the backing up and the going forward, and all the prescribed flows, all need to be tested.

Another thing that drives us *nuts* is grammar. When we put these fills in, we'll have a fill for "his"/"her" and a fill for "was"/"were", and all these other things, and you have to make sure that they all get triggered appropriately, *together*. And every now and then you'll be testing along and the question will read that "he were" or "we was" something-or-other and you just have to look for it, because you can't find it any other way. Now, maybe some of these [grammar] checkers—if they could be processed on every variant of the instruments, maybe that would work for us, I don't know. But it's a real problem, and it's one of those things that will really trip up an FR. If they're just reading what's on there and all of a sudden the grammar is bad that's a really big deal to them; they're going to see that a lot quicker than they're going to see a fundamental flaw in the logic.

We have to test the content—are we collecting what we set out to collect? Do we have it phrased the way we intend it to? That's another aspect of the fills: whether or not we have them coming in appropriately.

Two things that are more important now than they used to be—back when we were doing character-based instruments in DOS, these weren't quite a big issue. But now, in the Windows environment, the layout of the screen becomes important, and usability issues become very important. So we've added to our burden—finding some time for usability testing to test out the screen design and make sure that the choice of colors and how we do our buttons and everything is going to work well in the field.

So who do we have available to do testing? Well, we have the people who developed the systems get people to test the systems. The people who program the instrument—whom we typically call "authors"—have some level of testing that they do themselves. We have the survey managers, which I use generically as this group of people who are kind of running the trains, making sure that the instrument is properly developed. We have the analysts, the ones who asked for particular content to be collected—they need to see if it's being collected. We have the survey methodologists, the cognitive researchers, who are trying to figure out how to phrase the questions—they need to look to make sure that things are getting properly phrased. We often use our field staff for testing our instruments—once we have an instrument that's pretty much working, they're excellent sources for finding things like the grammatical problems. "Oh, I remember interviewing this nice lady and she had this characteristic but that's not really on this screen, so we really ought to fix it"—really good input. And, then, anybody else we can find ...*literally*. And the better we document the instrument, the better we can use people off the street to test it. If we can't document the instrument, then we're stuck testing it with the same people who are trying to get it produced and get it in the field. Which is where we kind of get a resource problem.

What tools do we have available to do our testing? We can review the code, literally, just read it from top-to-bottom, which is fairly easy to do in CASES. It's not quite as easy to do in Blaise, because Blaise has it split up into different components and it's a little harder to read in a linear fashion.

We can break the instrument into smaller pieces and test it a little bit at a time, so you can get your hands around it—I can just test my labor force questions independent of all this other stuff.

The way we do our testing primarily is that we manually simulate cases. We sit there and we bang away at the instrument, and we just invent these cases, and we just work through the systems and the instrument. And just try to anticipate everything that's possible.

We have tools to replay the instrument, so that if we've executed it one time on a particular case and then we want to go back and see how that worked in the field we can do that. We have tools to analyze the instrument that are the same as those used to document the instrument. We have, at the Census Bureau, a "Tester's Menu" so we can basically post these test instruments and have one of those menu options for processing the output so you can more easily check the output.[12] We have tracking systems that keep track of where the code is and what version of the code you've got, and that sort of thing.

We also have a testing checklist that I thought I would summarize a little bit for you. Basically, a manager can sit down and make sure that everything has been taken care of; they can make sure that the input files have been tested. They can make sure that they've tested all the substantive paths, with and without back-ups and with and without making changes to the instrument as you go along. We have instructions to make sure you've checked the minor substantive paths—again, with and without back-ups. [We're] also looking for questions that end abruptly or questions that have no people to ask them of because there's no path that gets you there.

Testing the fills—very big deal. Also we need to test the behind-the-scene edits; we need to test whether or not interrupting the instrument works. The outcome codes that Bill was talking about—we need to make sure they're functioning. We need to make sure that the systems and the instrument are properly talking to each other. We need to check the spelling and grammar. We need to make sure that the fields are being entered properly, and—as I talked about before—whether they're showing up correctly in the instrument output.

So, with all that, what do we need? Right now, we can't really automate our scenarios. I had an understanding that [the Research Triangle Institute (RTI)] might be somewhere on the verge of having something like that but I have not actually seen it. Our experience is that when you are in development mode the instrument is not stable enough to be able to generate the scenarios to just re-run this thing over and over again.

We can't really identify all the possible paths. If we can't identify them, how are we going to test them? There's basically the problem that the volume is too high—we just simply can't get to testing all paths. All the infinite possibilities, with the back-ups and a lot more possibilities for movement around the screen—right now, it's just a very time-consuming linear process. We do *A*, then we do *B*, then we go back and re-do *A*, and then we do *B*, and then we do *C*, and then we go back and do *A*, then *B*, then *C*, and then we do *D*, and *A*, and *B*, and *C*, and *D*—it's

---

[12]The Tester's Menu is a Census Bureau-developed software package for Microsoft Windows platforms that is intended to process and test instruments coded in Blaise or CASES. It is described in more detail in Dyer and Soper (2001).

about the only way we can get it done.

And basically we need to re-test the entire system after every change. Because, sure enough, if we don't, we'll introduce a bug that we didn't think we did, and it'll destroy a path—somewhere out there in the instrument where we didn't look—and we won't find out about it until after we've collected the data.

Our systems testing is basically distinct from our instrument testing, which is part of another linear process. We have to be pretty much finished with the instrument well in advance of fielding so that we can do an intensive amount of systems testing. Ideally, our systems people want our instruments done two months—completely certified, finished—two months in advance of fielding. I don't think we've ever quite succeeded, but that's the goal—that's so that they can feel comfortable that they've got the instrument fully functioning.

I've talked about the fact that you don't always have on your output file what you put in. That's another thing that makes the testing difficult. I've also mentioned that in Blaise it's more difficult to read the code from top to bottom. It's also difficult in the case Mark [Pierzchala] mentioned, when you have foreign languages—depending on how the translations are embedded or not embedded in the code. If you're doing a translation as you're trying to read along the code, and you only know one of those languages, it's going to get really complicated to read the code. Of course, if you know two languages, it's still probably going to be hard to read the code.

The systems are not really self-documenting, and that's what really restrains us in terms of resources to test.

OK, more deficiencies ... I wonder how many more I've got? Particularly in the case of [graphical user interfaces (GUI)], we need to test on all platforms. Once we've tested on our machines at the office, that doesn't necessarily mean it's going to work like we think it is when we get it on a laptop in the field. So we've got to test on all those. We have to look at performance on several platforms as well.

We never have enough time to test. We never have enough time to document. We never have enough time to do what we want to do. But particularly the testing gets scrunched at the end. Partly that's because we're constantly refining it—right up until we go into the field, we'll get a phone call that we really need to change this question. It's critical; we just passed this law yesterday that says we have to do x and you really have to change this question. We're just about to run with the instrument into the field, and that's a really dangerous situation to be in. It's critical to update the substance but—if you do it on the fly—you run the risk that you break something else and create a problem. And because the instrument is continuously refined you can't get a fixed set of test decks that you can use repeatedly.

I'm about worn out—did I run out of time?

CORK: We're right on time, so if there [are] one or two questions we could take them now and then go into a short break before getting into a computer science flavor. If there are any questions?

DIPPO: I notice that you're focusing on documenting the "what" and not addressing the "why." For instance, in the [Current Population Survey (CPS)], we have the response category be two columns—we did it specifically so that these were the at-the-job and past-the-job entries. And then when we went to go from Micro-CATI to CASES and we wanted to do something, if there wasn't a documentation of why we did something and why it should be included in the next version, we wanted to have a specific reason.[13]

DOYLE: I absolutely agree with you. There's also not sufficient documentation of the results of the pre-testing that we do, to figure out that you really did want to do it for that reason. But, today, I'm only focusing—I take my documentation problems a little bit at a time. For this group, what we're trying to do is to get them to figure out a way just to document what we did do and we'll get another group in here to talk about documenting why we do it.

What we're trying to do in the context of routine quality profiles for our surveys is to start incorporating documentation like that, which says, "We tested these questions, we made these decisions. And this is the critical reason why we've done x, y, and z." It's part of the quality of the data itself, as opposed to the mechanical or computing problem of documenting this piece of software that we've built up.

BANKS: This is probably a bad notion, but could one think about doing instrument development and documentation on the fly? You start off with a simple instrument that you think does a good job of capturing most of what you care about, you go out and use it—and one of your respondents says, "Well, I guess I have that sort of income; I'm going to receive a trust after 30, and the trust money gets compounded with interest. Does that count or not?" And nobody has a clue. So what the FR could do is to write down the narrative account of what that issue is, feed it back to the top of the system, and then you make a judgment call as to whether that's what you mean by "interest income" or not.

DOYLE: What do you think, Susan, can we do that?

---

[13] Micro-CATI was an early PC-based software package for computer-assisted telephone interviewing. It was used, for example, to conduct part of the National Crime Victimization Survey (NCVS) from 1989 until 1999 (National Research Council, 2000).

SCHECHTER: Well, I would say this. [For] the review process OMB wants to see as many of those scenarios as possible, from those reviews of the instrument, since all of that influences burden. So we're interested in those observations from the field. But I don't know how that would solve the documentation problem; it's more of an enhancement to the process.

BANKS: I think it's a part of the documentation; on top of the original, you have an "excuse list" explaining why a change got made.

GROVES: A completely different question: on your slides on testing, you had a role for methods researchers and for analysts. Is that a role that goes beyond just checking that the specs they delivered are being executed?

DOYLE: That's correct; there is a role beyond that.

GROVES: What is it? Are you letting them re-design in the middle of the go?

DOYLE: In the course of developing, our experience has been that we start out and say, "Do this question." And we implement the question. And then we start to test it and find that, "Oooooo, this isn't quite working like I thought it would." So let's go back and change the specification so it works slightly differently. It also may happen that we're implementing the automated instrument at the same time that we're doing cognitive testing, so that while we're doing cognitive testing and find that these words don't mean exactly what we thought they would, let's change them to these other words. And you have to go back and embed it in there. So some of it is an overlap, to try to reduce the overall length of time it takes to do all of this work.

MANNERS: Just a comment; I think it's important, to follow up on Cathy's question, to build in your documentation plans while you're planning and building the instrument, and to build in links to external databases that give rationale and reasons for items.

DOYLE: Absolutely. And one of the benefits that we derive from using the database management systems to develop specifications is that they are bigger than just the specs for the instrument—they have instructions for post-collection processing, so you can tie in that work. And eventually they're going to be able to accept back in statistics from the fielding, so you get all your frequency counts. There [are] also ways to link back to other documents that provide the research for, say, the cognitive testing for question *x*. It's more than just solving the instrument documentation problem.

PARTICIPANT: I think that one thing we seem to be increasingly coming to [is that] at the same time that we're trying to get better at enforcing having all the specs, and having better specification, I think it may be time to realize that the instrument itself is becoming part of the design border ...

DOYLE: It's become what?

PARTICIPANT: Part of the design border, the design desk for survey researchers. And that they cannot fully think about the specification and design of the instrument almost until the thing's half-programmed. And while we don't like that from the programming side, it may be really inevitable. And we need to start thinking about these specification database systems that are starting to roll out as somehow being a reasonable workspace for a non-programmer, for a survey researcher, and actually some creative work can be done inside them.

DOYLE: Yes, excellent.

PARTICIPANT: And just finally accept that it's going to be iterative throughout the whole process.

DOYLE: Yes ... can we keep going?

CORK: Actually, we've got a lot to get through, and I want to keep as much on-track as we can. I hate cutting off discussion, but let's take a short break and then we'll start up again at 10:40.

[A short break was held.]

SOFTWARE ENGINEERING—THE WAY TO BE

Jesse Poore

CORK: This morning, we had an overview of current practices as laid out by the Census Bureau. We're now going to turn toward the computer science perspective, and introduce Jesse Poore, who is a professor of computer science at the University of Tennessee.

POORE: You're probably wondering what a person like me is doing at a place, a meeting like this. I don't know the answer either, but I have some data. I was giving a speech at a defense workshop a few months ago and described myself as a drab and humorless person. And immediately after that Mike Cohen ran up and invited me to speak here! *[laughter]* So, I don't know if there's a connection or not.

I want to say just a few words about what I understand your mission to be, and a few words about what I understand your situation to be with respect to software. I'm here to talk about software. I want to remind you about Moore's Law, which addresses the change in technology. And then I'll talk about software engineering, which is the primary reason I'm here.

I know that what you do is extremely important. You are responsible, probably, for the data that's behind [Alan] Greenspan's most exciting speeches. And I know we have redistricting every ten years which is really exciting. And there are housing starts. I built a house once, and I always wonder when I hear about the housing starts if they're going to finish any of them. [*laughter*] The punch list is always pretty long.

Seriously, I know you're supporting a lot of public policy decisions in all branches of government, and the Bureau of Labor Statistics, the Bureau of the Census, and others have a very distinguished history. And in fact they can be credited with a lot of the early progress in computing and data processing. But the question is really: what about the future? And when I think about the future of the business you're in, it seems to me that the public interest groups, the news groups, the special interest groups are really encroaching on your turf. I think they're moving in on you. And I was thinking about how can you win in this business, and I think the way to win is to be first with the right information. The other guys don't have to have the right information; they just have to be first. But you have to be first with the right information. And you have to have trust associated with your data.

So I look at the problem as how do you move fast—how do you move really fast, so that the decision-makers will turn to you for information instead of turning to the person who's handiest? And how are you first with the right information, that's statistically correct, secure, and trusted?

So I don't think of your job so much as the design of surveys and all the problems associated with survey design, as we were just talking about. I look at it more from the point of view of the business you're in: who your competitors or threats might be and what you have to do to win. And the role that software might play in that.

I spent some time looking at your software situation—not a lot of time, less than a day.[14] And my conclusion is that there's nothing unique there or overly complex; I just don't see any hard software problems. Now, I'm beginning to think based on the earlier discussion that there's some merging of software issues with survey issues. And so maybe what I'm looking at is just the software problem, and really it's the problem of the survey designs being munged together with some of the software problems. So, to the extent that that's the case, just keep in mind as I'm talking that I'm thinking mostly of the software problem, and you might want to put the survey design problem on a separate plane. And it might be very similar; it might be an analogous problem. But it would be important to keep the two separated.

But, even though I don't see anything unique or overly complex, it's complex enough—it's not an easy, casual problem. So you have to take

---

[14]This is a reference to Poore's participation in a planning session for the workshop, which was held on December 11, 2001.

a very structured approach to the software engineering, and I think you have to pay special attention to the role of the technology in what you're doing.

Moore's Law says that basically you get a doubling in the bang-for-the-buck every 18 months.[15] And it seems to me that this is extremely important for you, that you have to operate from the point of view that you're going to be using very modern technology *all* the time. And I've heard DOS mentioned once here today, and I've heard other things mentioned that seem to be throwbacks to an earlier data processing era. And I think you have to be more modern; you have to be very modern all the time.

One thing that Moore's Law doesn't really cover directly—but is implied from it—is the fact that you get new dimensions in this technology. The fact that things get faster-faster-faster, cheaper-cheaper-cheaper, smaller-smaller-smaller means that you can get new technology that you didn't get before. Things that you can carry in a laptop that once would have filled a room—the wireless technologies and all of the multimedia. Like I said, I think that the news groups and special interest groups are in some sense your competitors. And they're out there using the media in various ways—wireless, broadcast, and so on—probably even having their samples drawn before they know what the survey will be.

So I think you take into account all of these technologies that might be available to you. And in order to reap the benefit of Moore's Law, you have to stay in the mainstream of software. This is easier said than done, but if you end up in some backwater and new hardware comes along, new software comes along, new operating systems, new applications, and you can't get out of the hole you're in, that's an especially hard problem.

There are two sides to this. Unfortunately, it isn't always the case that the first technology to the market becomes the standard. Sometimes it's the second one, or the third one. So you've got to be very modern, but you don't necessarily want to be the first to jump—you want to be the second to jump. Just hold back a respectable distance there, but stay very, very, very modern.

Now, one idea I want you to think about is product line architecture. This [slide] is *not* an architecture for your product line [*See Figure II-2*]; this is a picture to let me talk about product line architectures, because I don't know enough about your business to even pretend to give you

---

[15] Specifically, Moore's Law is the popular term used to describe Intel co-founder Gordon Moore's 1965 observation that integrated circuit density—or the amount of information storable on a given silicon chip—doubles in magnitude every year. This rate of increase was later adjusted to a doubling every 18 months. As it has come to be known colloquially, Moore's Law is a general gauge of the rate at which computing power has grown over time.

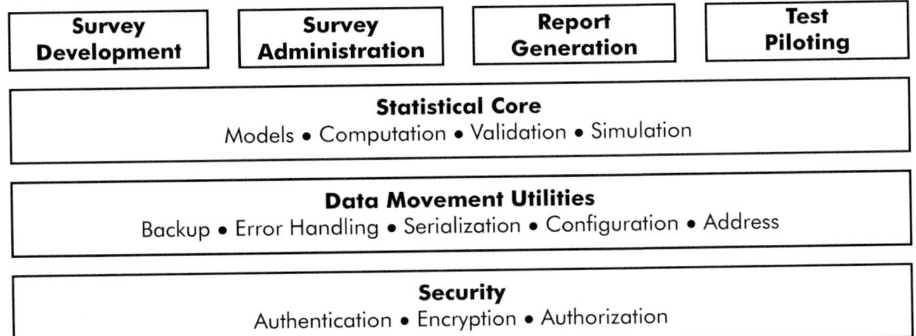

Figure II-2   General structure of a product line architecture.

SOURCE: Workshop presentation by Jesse Poore.

an architecture either for your software or for your survey development. But the concept is this. Product line architectures exist for companies like Hewlett-Packard, which makes laser printers. And they roll out new laser printers every 12, 18 months, something like that—each one of them is a variation on a theme. Each one contains some new content, but a lot of existing content. Nokia makes telephones—they roll out one telephone after the other. Each telephone has something new about it, but contains a lot of old content. And so you might think of your own software systems or your own surveys as being part of a product line, where each one that comes out is a variation on the theme—it's got something new, but hopefully it's about 80 percent old, familiar, working stuff.

So the concept of the product line architecture is to separate various concerns. In your case, I was thinking security might be one of them. You would want to separate out all aspects of security from every laptop you're using, from every desktop machine in your organization—all the way through your big data centers that process a lot of information. You would like to have security cut across all of that, so that when you change security in one piece you can update it throughout the systems. Meanwhile, you want data movement to be totally separate from the security issue, but likewise data movement should be a matter that's comprehensive across all the platforms, all the laptops right on into the data center. The statistical analysis core of your systems shouldn't in any way be intertwined with your data movement utilities. When I hear comments like, "When I make a change over here it breaks something over there," it makes me think that you don't have an architecture in place, and that there's not a sufficient isolation—separation—of concerns in the design.

So the concept is to have this overarching architecture that helps you to separate concerns and make everything that you do more modular and more easily dealt with a piece at a time. You don't ever want to have to change your whole scheme; you only want to change a piece at a time.

The other thing is to consider an incremental development strategy where—with this architecture—you would have some multi-year series of increments. You've got to be planning and anticipating the technology—you may be wrong but you should always have a working hypothesis of where it is and where it's going. And you update that all of the time. So you have a rationale for this series of increments—what you're going to change, when you're going to change it, and why you're going to change it. You should constantly reinforce this architecture or—if you find that the architecture isn't supporting something you need—you should change the architecture. You want to work in the smallest natural increments; you want to get away from taking on these huge challenges. I once was working with Ericsson—I do a lot of work with companies, so even though I'm the pointy-headed academic, most of what I'm saying is in use in a lot of companies. Someone in Ericsson was planning a project that was being pitched to a vice-president in Montreal, and they came up with something like one-half million staff hours, a half million labor hours for the project. And the guy says, "We've never had—in the history of Ericsson—a project that was bigger than about 200,000 staff hours that ever succeeded. So, no, I don't approve of this project." [*laughter*]

Even that's pretty big; everything's relative. In my research group, my small laboratory group, we look at everything in terms of three or four people working on something for three or four months, and we want an end result. We want to be in and out with a finished product in a short amount of time, relatively small number of people. I think that's extremely important, and my idea of a small increment—the smallest natural increment—is probably a lot smaller than you would think about. Everything has to be properly staffed and scheduled, and you want to meet these schedules and budgets with quality.

The idea of incremental development is that you don't take on a project—a big project—that's going to be, you know, 20 percent complete, then 40 percent complete, then 60 percent complete, then 80 percent complete, and 80 percent complete, and 80 percent complete. [*laughter*] You don't want to do that. You want to have a piece of the job where you can say, "This 20 percent of the job is 100 percent complete; it works, I can demonstrate it to the end user, to the customer. It may not do much, but it's 100 percent complete." And then you do the next piece, and you add on to that. And, of course, planning the increments is critical because you want to do the most important things first. And then your 40 percent of the project is 100 percent complete. Absolutely done, in every respect—documented, tested, ready to ship, if it's worth shipping. And then you move on through until finally you ship the final product. This creates a much better work environment than taking on the whole big project and farming out the pieces, and then trying to get them to come together in an "integration crunch," as it's called. If you do incremental development, every cycle—every increment—is really an integration test, an integration of the product up to that stage. It works very, very well.

You want to create an environment of success. There are too many data processing organizations that are just living year after year in failure. They never do anything on time, on schedule, with the quality. It's just failure, a constant life of failure. That's no fun. So it's much, much better to create an environment of success. And to do that you want to have all your standards, conventions, and styles documented, carefully selected, carefully reasoned. I'm not advocating any particular standards, conventions, or styles—I'm just saying you should have them. And that they should be well documented, well followed.

There [are] lots and lots of basic tools out there to support standards, everything from project planning to configuration management and so on. I also advocate that you work in teams with peer review, so that no work product no matter what it is—specification, code, test plan, anything—every work product is only done when three smart people think it's right. So that everything is visible, all the plans are visible, all assignments, all the progress, all the problems are visible—they're out there, and everyone's talking about it. Which means that you manage to what I call "inch pebbles" rather than milestones. The milestones are great, but I want to know what's going to get done today—what's going to get done this week. Don't tell me about three months from now—I want to know about this week, maybe today, maybe tomorrow, what's going to get done.

You want to avoid this game software developers play, called "schedule chicken." [*laughter*] The way "schedule chicken" works is that you have, say, two groups in an organization—they're both behind. But each one of them thinks, "I'm better off than the other guy." And each one of them thinks, "The other guy is going to feel the pressure first. That guy will break. I'm tough, and I'm going to manage it through." So they both keep lying about their progress, at every review, at every meeting— hoping the other guy will break because I know if the other guy breaks and gets some more time in his schedule, then I'll announce that, "Hey, since there's a little more time I can do a little extra here and catch up." And that's the way you wind up with these projects that are 80 percent

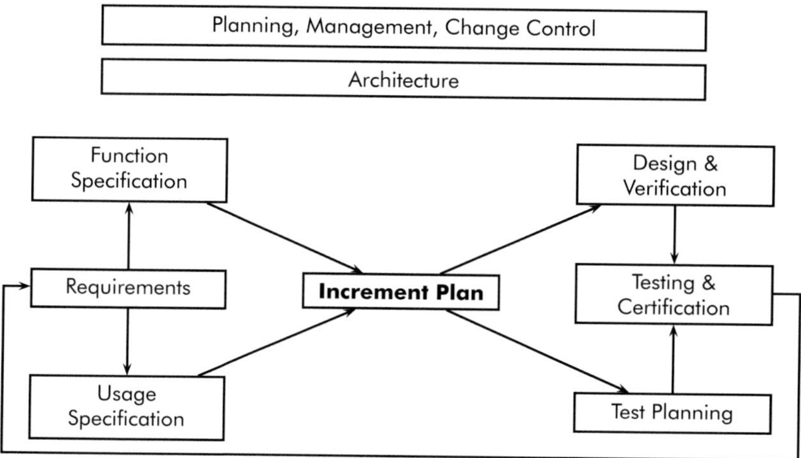

Figure II-3 Schematic model of a successful software environment.

SOURCE: Workshop presentation by Jesse Poore.

done, 80 percent done, 80 percent done, is with "schedule chicken." So we don't want to allow that. So the inch pebbles will help solve that problem.

So the picture might look something like this. [*See Figure II-3.*] Planning, management, change control—all of that is ever-present throughout the project. It requires eternal vigilance. It should never go away, it should never become slack—it's always there, and you talk about it every day. The architecture is something that should be fairly stable. Again, it's always present—it might change, but it's changing less rapidly than other things. Then you go into the development activities where you have your requirements. And I like to think of the requirements in two ways—work with them first as a functional specification, and then also work with them as a usage specification. So you're talking about not only what's going to be there and what it's going to do: you talk about how it's going to be used. And how the thing's going to be used should always be right up front and ever-present. I looked at a new system that was going into the Water Management District of Southern California, which could also be called the Los Angeles Water Company. And they were putting in a big new system to replace this mainframe with slave terminals attached to it—a pretty efficient scheme. So they asked me to take a look at it and see whether it was going to come up on schedule, work on time. The first thing I said was, "Well, let's see this person who works at this counter log in and use the system." It took 27 steps to log in. And I said I didn't think it would work. [*laughter*] If every worker in the whole water company has to go through 27 steps just to log in, when the old system had three, then someone hasn't been paying very much attention to the usage requirements.

Well, anyway, you look at the functional activity, the usage activity; think about the increment plan, the smallest natural increments. Then you branch and go into design and verification of the code on one hand, test planning on the other hand. When the code's done, the test planning's ready, you put them together and do the testing and certification. And these little layers are to indicate multiple increments, and you cycle back—reconsider the requirements and see if everything is still the way it's supposed to be. If it is, go forward; if it's not, you make the changes and you go through the next increment. So you operate in that sort of software development cycle within the framework of the architecture, in the context of the architecture and the ongoing management, planning, and change control.

I guess I don't mean the same thing by "documentation" as was discussed earlier—maybe, maybe not—but I do have that attitude that if documentation is important at all, then do it first. If it's not important, then don't do it at all, I guess is the answer. But here I'm talking about documentation in terms of software development, software engineering. And so we want to maintain written requirements. You want to convert those requirements to precise specifications—that's probably *the* most important work in the entire activity of software development, getting those loose English statements in various memos, documents, sometimes a booklet if you're lucky, but those statements of what's presumed to be wanted—convert it into precise specifications. Something that's tight enough that programmers can do their job without having to invent or to make up information about what was wanted. The programmers should not have to make any decisions about what the end user wanted. And so you try to get those into the specifications first.

In doing the process of doing that, you will invariably find conflicting information; you'll find missing information. Things are inconsistent and you have to make decisions. You tag and trace all those decisions so that later on when you find that a statistical algorithm was changed, it was changed because a statistician told you to change it—not because, say, a field interviewer told you to change it.

Prepare the user guides while you're writing the specs; prepare the test plans while you're writing the specs. The hope is that you will not design an untestable system. There's no reason to design untestable systems. You'd like to always know how you're going to test the system, that you're going to be able to test it in a satisfactory way given the time and money available. Then you produce these complete, consistent, and traceably-correct specifications—that's what I mean by documentation. All of that is documented from the top down, and then the programmers can take over and write some really good code. Yes?

BANKS: In the software world, when you try to lay out your requirements, it's pretty well understood that you want the software to do particular things.

POORE: Sometimes.

BANKS: Well, OK. But my sense is that we have a less clear understanding of what specific legislative decisions, or judicial decisions, or executive decisions may be driven by the results of the survey.

POORE: That may well be, but I would have to ask: how big a factor is that? Is that 50 percent of everything that's going on, or is it 1 percent? And, if it's 1 percent of the trouble, I'm willing to let you have 1 percent of the slack for that. But I won't give you 50 percent of the slack if it's 1 percent of the problem, right? So those things happen.

I've been trying to think of an application that I've done that's similar to survey data—just looking at my own experience for things—and I can't come up with anything that's right on. But the closest I've come up with is this: the Oak Ridge National Laboratory tracks a lot of toxic, hazardous waste material, everything from five-gallon buckets, to drums and barrels. Some of it is liquid, some powder, and so on. Every container of it has to be tracked—it's put on tracks, it's put in storage, it's put on railcars, it's moved here, it's moved there. And the whole idea is that none of it ends up in a public school cafeteria. [*laughter*] So that would be bad. So they had these survey forms; they had these laptops. They go to the people who are containerizing it at the source and putting it on shipments. They'll go interview people who are receiving stuff and see what they received. And they'll cross-check what was shipped with what was received. They'll ask the truck driver who did so-and-so, you know, "You went here ... No, you didn't go up and dump anything in that lake?" It's just all these different sources asking question after question, and cross-checking and correlating everything to look out for a problem. When I first looked at this, I did some estimates of how many paths there would be. And I said that it's untestable—you don't have enough time or money to test it. And they said, "Fine, so what do we do about it?" And I said, "Well, you reduce options. You cut out gratuitous complexity. You get rid of questions that really don't matter, and you narrow the thing down to the things that are absolutely important, and you control everything. And so we were able to get a system that was testable, to test it and get it into use.

I was talking with Daryl [Pregibon] earlier about the statistical analysis that follows from some of these activities, and I don't understand how the analysis at the other end can make use of a lot of the detail that it sounds like is going into these questions. But I'm way off track.

Anyway, you don't have 100 percent control. But you control absolutely everything you *can* control, and isolate that other part—it makes a big difference.

Well, you're going to hear a lot about model-based testing, so I won't talk too much about that. I do think it's important. Well, your quality goals and requirements are going to be part of the architecture. There are ways to meet those quality goals, and I think that model-based testing is a good approach. With the kind of complexity I'm hearing about here, I'm convinced that you will have to automate that. If you don't automate your testing, it's unlikely that you will do enough testing of sufficient variety in a short enough time at a low enough cost to satisfy you. So you should look into automated testing.

I would say you develop the code last—the coding is the easiest part, particularly when you have good specifications. Coding is very, very important because you want it to be done right, you want it to be as simple as possible, clear, transparent, easy-to-read, no tricks, and to exactly what the specification says it should do and nothing more. And the way we like to work is that when code is written we verify it with the specs and all the standards and all the interfaces; that's a labor-intensive process that involves at least three people. We check every line of code in detail against, again, all the standards and all the specs. It turns out to be very economical in the long run because it eliminates so much re-work that you more than pay for it.

Now, it sounds like you're doing a similar thing with the survey development, and I think that's probably the right thing to do. I think you'll do more good with peer review of your surveys than you will in testing your surveys, particularly if you develop some nice modular structure for putting them together.

When you were talking about all the paths through your surveys, I was thinking, "Well, there are paths through code." And, sure, there's an infinite number of paths. But you don't deal with that. When you're dealing with code you deal with a finite set of structures—a finite set of units—and you argue with respect to the finite set of units and not the infinite set of paths. And a similar thing should be possible with surveys, I would think.

I was asked to talk about testing, more so than other aspects. And I will continue to talk about that some. However, my view is that you've got a testing problem secondary to having a specification problem. So my view is that you should really focus on your requirements and specifications first, and that your testing problem will get a lot easier, a lot better.

We build models for software testing, and when we build these models we perform what we call the "crafted" tests first. Those are the ones required by contract or law or industry standards—things that you specifically want to test. We record all the information from the tests when we're running the tests; we monitor the progress, monitor stopping criteria. The people I work with are generally testing in order to establish or assert a certain reliability to the code, so they're monitoring the progress and stopping conditions to see when they've collected enough empirical data to make those sorts of claims. If the progress is good, we keep going; if the progress is bad, we can see that the code isn't going to pass and you return it to development. If you're meeting your goals, you can move forward. Yes?

PARTICIPANT: Is it typical for the testing group to be independent, i.e., be independent of the group that's writing the specification or authoring the instrument or doing the programming, so that they have their own workload? Rather than adding that onto the responsibility that they already have and having specification writers coming back after it's been authored and doing re-testing? Is it typical for a testing group to have that separate focus?

POORE: Well, you can find most any organization imaginable, so there [are] lots of different ways of doing things. I would say the more mature organizations tend to have a separate testing and quality control group from the development group. The people I work with tend to get those groups together at the early stages to talk about the product and what they're developing and how they're developing it. And then the coders—the programmers—go off with the specification in one direction, the testers go off with the specification in the other direction, and they work independently until the two come together.

But I certainly know of organizations that just aren't big enough to do that, so the developers also do the testing. And in those cases we try to build some independence, state the criteria up front as to how the product will be tested and hold to that statement.

After the products are released, I would recommend that you track all field-reported failures and understand why those failures happened, look at all the factors of production. I mean, when you do [this] kind of work, you're doing it in an environment where the failures are relatively few. If you're working in an environment where you're just overwhelmed with failures it doesn't make sense to do a lot of these things. You do a root-cause analysis on each and every problem and find that it's the same thing—that people didn't know what they were working on, or didn't understand the programming language they were writing in. And you should find that maybe once—not a dozen times or a thousand times. And when you track field-reported errors in an environment of

$ × 1 in Requirements Analysis

$ × 2 in Preliminary Design

$ × 4 in Detailed Design

$ × 8 in Code and Test

$ × 16 in Integration Test

$ × 32 in Field

Figure II-4 Conjectured multipliers on cost of correcting errors at different phases of a software design project.

SOURCE: Workshop presentation by Jesse Poore.

success—in an organization that is typically doing good work, meeting schedules, budgets, and quality—then you're doing this from the standpoint of finding out what needs to be revised about the architecture or the planning or your process or your technology or the staff, in order to do better and not to repeat those kinds of problems.

The data look sort of like this. [*See Figure II-4.*] You may not be able to read that, but number one back there is requirements analysis, number two is preliminary design. One of the groups I work with a lot—Raytheon—uses this terminology: requirements analysis, preliminary design, detailed design, code and test, and so on. And their data is not exactly a doubling; it's close. In some of the phases it's a doubling, in some others it's just a little bit below that, as to what it costs to correct an error. If you find and correct an error in requirements analysis, then it is unit cost. If that error escapes from requirements analysis and doesn't get caught until preliminary design, then it'll cost you twice as much. If you don't even catch it there and it escapes all the way to detail design, you're now four times as much as it would have been before. And if you look at some mature products in good organizations ... fixing field-reported errors can be about 32 times the cost of fixing an error at the outset. So fixing an error might cost ten or twelve thousand dollars. And they're getting two or three hundred a month. And that's steady-state. Do a little bit of multiplication there, and you come up with some big numbers. That's the cost of field-reported errors. And so you can see the motivation is very strong to prevent those. Yes, sir?

PARTICIPANT: I love your slide, I love your point. My experience has been that the steepness of the curve is much stronger than that, and that the cost of an error in the field is much more than 32 times.

POORE: Right. I'm emphasizing again that we're talking about what you might call CMM Level 4 organizations, who have methods in place for doing all of these things.[16] If you're less mature as an organization, then I guess it would be much more costly.

At any rate, my point is that a lot can be denied in the software development world; the one thing that can not be denied is a field-reported failure. You just can not sidestep that. So, even though everything we know from theoretical considerations and experience and so on says build the quality in early, do the right thing at the very first stage—there's no way to get a handle on that within a software development organization. It's very difficult to jump into the middle of detailed design and say, "You guys are not doing a good enough job, and you're going to mess things up at the next phase." That's hard to do. That's why we're always finding ourselves starting at the bottom of the food chain; you can not deny the field-reported error, or its cost, and you start moving upstream. And you say, well, we can't blame all this on integration and test. You can't say that they should have caught it. That's just an oversimplification. Maybe they should have caught something; you fix whatever they should have caught. But then the next problem is what to do with the things that they had no way of catching, they should not have caught, that [are] unreasonable for them to catch. Well, "they should have been caught earlier." And so it goes—it's sort of like a little proof-by-induction or something, as you go all the way back to the base. And the extent [to which] you get it right in requirements analysis and preliminary design will determine everything about your success in terms of schedule and budget from there on out. Yes?

DOYLE: What do you do if you're in a circumstance where you need to get to started but you really don't have the scope laid out to the point where you can do a complete requirements analysis?

POORE: Well, I don't accept those terms. [*laughter*] I don't accept those terms. Why do you have to get started? What is it that you're getting started to do? I mean, how are you getting motivated to get started if you don't know what the job is yet?

DOYLE: Well, what if you know one-third of the job, pieces here and there?

POORE: Ah, OK—then you have to have this architecture that will let you do productive work on what you know is available to do, and stop there, so that you won't do more work than you can keep. You

---

[16]The Capability Maturity Model (CMM) for Software is a scale for assessing where software organizations stand in the evolutionary process from chaotic early practice to fully mature and disciplined development. The CMM rating scale is developed and revised by the Software Engineering Institute at Carnegie Mellon University. In increasing order, the levels are: (1) Initial, (2) Repeatable, (3) Defined, (4) Managed, and (5) Optimizing.

don't want to do work that you throw away; you don't want to do re-work.

DOYLE: But what if the parts that we know don't follow good, sensible, modular design at all?

POORE: Then the chances of you doing any productive work on it that you get to keep are nil. So work on something you do know. You can't have it both ways, is my point [*laughter*]. You either have to have an . . .

DOYLE: Our situation is that you can't wait because you'd run out of time, waiting for every "i" to be dotted and "t" to be crossed.

POORE: But I'd say you just can't have it both ways. You either have to have a structure, an architecture, an environment where you can do work and not lose it, or there's no point in doing it. If it's going to result in re-work, if it's going to be thrown away, then . . . Let me give you the best example. I never succeed in convincing people of this until after they've tried it; can't win it on argument. Unit testing—it's a big waste. Because by definition when you do unit testing and then you try to integrate and things don't work, you have to go back and make changes. And when you make those changes you're undoing a lot of the unit testing that you were doing. So that's an activity that's very, very common in software development, awfully hard to change. But when you get any organization to get onto an incremental development instead of a unit separate, unit test, and integrate—they'll never go back to it, and they'll cut out that wasteful step. But it's a hard argument to make.

TUCKER: From your earlier description about dividing into small increments, you put a heavy emphasis on project management.

POORE: Yes.

TUCKER: And I think we have a real problem in trying to have enough of that talent to do the instrument. What you're talking about is day-to-day, week-to-week intensive . . .

POORE: Well, there's a little bit more to it than that. The management is important, but incremental development forces a different ordering of events.

For example, I worked with the IBM Storage Systems Division in Tucson. They make tape drives, disk drives, disk storage units, and things like that. The first time we did incremental development there—incremental development says that you've always got to have something useable, deliverable to the customer, from the user point of view. And so that would mean that the first thing you'd have to do is the power-on reset. You know, turn the tape drive on and turn it off. And they said that we always do that last—there's nothing interesting or hard about that, so we do that last. So I said, "Well, we can't do our increment test unless you can turn the thing on and off." So, OK, they were going to try the method *our* way, and they said we'll do it. They normally would have done it last, and they did it first. So the first increment, you could turn the tape drive on, it would go through its internal test, come up ready to work, and you turn it off, and that's all it would do. But if anyone wanted that tape drive we could ship it; it would work.

PARTICIPANT: I don't want it!

POORE: Oh, you've got one of those! [*laughter*]

The next increment we put in the fundamental commands to move the tape forward and backward. Turn it on, move the tape, turn it off—that's what it would do. The next increment we put in the basic read-and-write commands, and we could read and write data. At that point, the thing was actually useable. The next level, we implemented more of the SCSI protocol for that tape drive, and at *that* point it was useable and could have been shipped. And in the final increment we put in the final bells and whistles of the SCSI protocol, which are probably rarely if ever used. So that at any point along the line you could quit.

So for your survey instruments, I would like to think that each one is a variation on a previous theme, that you could find one out there that's very similar to the one you're doing. OK, you say no. Now, let me ask you this: is the question you're answering—is the purpose you're serving—is the reason for doing this survey in any way related to any other job you've been given before? I mean, is it similar?

DOYLE: Well, we often have a data series where we do some repetitions. And so in that case we have some similarities. But we have a lot of one-time surveys that are unique for a particular compilation.

POORE: But aren't the one-time surveys, in some way, similar to each other?

DOYLE: They're always very unique in the questions that they pose.

POORE: But ... I keep wanting to find a framework, to say that you could adopt this framework and go in and work with the details.

ONETO: Well, I think that in the survey world we really did embark on an incremental development strategy ...

DOYLE: Oh, yeah ...

ONETO: ...ten years ago, but we just didn't know it. And the increments have been expensive, because the increments are each time we've fielded a survey. And I think that there has been a building of knowledge and a building of expertise. And, according to what you've presented here, I think most of our maturity perhaps has been at the requirements and specifications development stage.

Prior to automation our biggest time constraint was getting OMB's clearance at the time we were getting ready to go into the field. But now we know we need a clear specification and requirement stage, a clear testing stage, and all this kind of thing.

So I think that we have made a lot of progress, and as we go from survey-to-survey we do borrow a lot of expertise. But I think that it's with the testing and the responding to field problems that we have major problems.

POORE: OK, one question and then I've got to finish up here with a couple of other slides.

PIERZCHALA: I just wanted to make the point that there are classes of surveys.

POORE: Right.

PIERZCHALA: There are variations on a theme; there [are] household surveys, economic surveys, social group surveys, health occupational surveys. So there are variations on a theme.

POORE: And it would seem to me that you could inherit some large percentage from some previous survey and work within that framework. But, like I said, I'm not in that business.

So, in conclusion, I think that the advice is manage, manage, manage. You've got to manage expectations. You can't have people expecting a job to be completed when the workers don't have any reasonable chance of making that date. And so you have to get realism into the thing early.

Manage the technology, manage the users, manage the requirements, manage the environment, manage the staff—and even manage your productivity rate. You should have a steady stream of product come out at a predictable quality level and a predictable rate.

Some of these ideas you can find in this book;[17] it's just one way of doing things, it's not the only way of doing things. And I will end on that commercial message. [*laughter*]

CORK: I don't want to be too much of a traffic cop, but we also want to be out of here before seven o'clock. So, hopefully, we'll have time to discuss some of the general themes Jesse raised later on in the afternoon and certainly lively discussion over lunch.

AUTOMATION AND FEDERAL STATISTICAL SURVEYS

Bob Groves

CORK: But, right now, bridging these two perspectives, we asked Bob Groves—who is the director of the Survey Research Center at the University of Michigan—to comment on the two presentations. Bob?

---

[17]The book referenced here is Prowell et al. (1999).

GROVES: This is not the first such meeting that I've attended. In fact, listening to these talks reminds me of the first one, which was in the mid-1970s at a beautiful resort near Berkeley. And the conclusions of that meeting, as I recall, were that we were on the cusp of a revolution. That we could totally automate all of the activities of survey research. And that, in fact, the lessons from then-extant computer science were easily applied to the task because it was a rather simple one.

I think what was missed in that meeting was that the attention was on the design of a questionnaire as a software kind of problem, and what turned out to be more difficult for the field was all the stuff *around*—the so-called systems stuff that Pat [Doyle] was talking about.

Then the other conclusion was that this would be a radical reduction in cost of collecting data for human and business populations, because it was so easy to change an instrument. It could actually be done at the very last moment; in fact, it could be done in the *middle* of the field data collection. We [would] really [be] able to completely revolutionize the timeline of development, which at that time people were fretting about.

That didn't happen either.

And then the final thing was that this should allow us to have stored archives of software that would be applicable to the kinds of questions we ask in all the surveys—because aren't these surveys similar to one other? And you could just take, say, the demographic questions and store them, and when Pat did it a few years later she could use exactly the same code. Well, what happened to that prediction?

The prediction was naïve in the sense that the demographic measures haven't stabilized. They change all the time; the code changes—the functions change.

And so we are where we are.

I have a few things to say, along the same lines. My job, I think, is to attempt—although I can't— to bring these two perspectives together. [You'll] find me speaking more to our computer science colleagues, I think, because I want to set the context. I think that's really important.

The federal government spends about $3 billion a year in statistical activities—that's not a lot of money, if you think about comparable commercial sectors. And, in fact, the commercial sector in terms of surveys and statistical activities is probably triple the size of that. This is a relatively small enterprise, even though we think of it as *our life* and everything in the world.

The surveys that are done here are of extremely long duration and are relatively stable . . . despite what Pat says. [*laughter*] So, I have colleagues who work at CBS News' survey unit who have two hours to put together a survey, that is done over a four-hour period, to be reported by Dan Rather the next day. The surveys we're talking about here are very, very different, and they might have developmental timelines where the questionnaire would be constructed over a 12-month period, or an 18-month period, and the questionnaire would remain stable for *years*. These are very different worlds and pose very different problems for software.

The federal government consists—on the survey side—of very large organizations that have a long history. They have work structures that are bureaucratic in nature by design—"bureaucratic" with a positive spin on that word. But these divisions of labor are very slow in changing, and much slower in changing than a commercial organization because the impact of external environmental changes is buffered by organizational requirements.

Finally, a distinct aspect of this that Pat touched on is that there is a devotion to providing data back to the people. So the product is not just Dan Rather saying the next night that 72 percent of the American public favor what Bush is doing—it's actually much more complex. It's large sets of data that need to be accessible by diverse users, and checked, and analyzed, and re-analyzed to check the credibility of the information. So these are burdens that aren't faced by others.

There is a long history of survey automation that you can just kind of go through here. Let me go through these terms on the back end here. "ACASI" is Automated Computer-Assisted Self-Interviewing. "TDE" is Touchtone Data Entry, and "VRE" is Voice Recognition Entry. We have, as a field, automated ... if you look at this, these automation efforts are five to ten years behind software and hardware development in other sectors. They lag. And one question central to a group like this is: is that lag good, or is it bad?

The problem of innovation in surveys deserves its own attention because a group like this would, if successful, innovate. And there's a problem in innovation in surveys. One is that a scientific orientation in the field is only in certain components of the field. It's certainly on some design aspects and some analysis aspects. But the survey field developed so that the middle—the questionnaire, the instrumentation, the collection of data—is designed as a relatively routine, stable-technology, production shop. Our problem is changing that, and it's hard to move those structures.

Historically, there's a fairly low investment in research and development for data collection and dissemination, relative to other sectors. This is not a group, this is not an industry that is planning for the future ten years out and has a set of R&D teams that are really planning what's going to happen. It doesn't look like the pharmaceutical industry, OK? And hence there are weak ties between R&D and production.

And a concern, more recently, is that the industry has relied on a relatively unskilled labor market for its data production activities. These are close-to-minimum-wage persons who are increasingly being given software and hardware products of rather large sophistication. Could that labor market change? Should we be designing for a labor market that becomes much more technically astute? It isn't clear. But certainly that's the current state.

The challenges, I think, of this workshop are the following. We do have to be real careful about definition of tools. So Michael [Cohen] was interrupting Pat from time to time saying, "define what you mean by that." I think we ought to keep doing that, and not let anyone get away without defining their terms if they're unknown to you.

There's a real problem with choosing the level of abstraction that we speak at. My belief is that the errors of the past, for gatherings like this that I have attended, is that we have allowed ourselves to talk at too abstract a level, and hence—we actually miscommunicated. Both sides were saying the right things, but it was irrelevant to the problem we were facing. And that means assessing the solution in the context of the users.

The solution set, I think, has to maintain attention to the fact that there isn't a well-defined answer to the question, "What is a survey?" So my friend at CBS News is a survey researcher. And Pat is a survey researcher. But they live completely different lives with regard to [their] issues in terms of design and future problems and innovation. And we really need to fix our attention at what we're trying to solve.

And, by and large, I think that the largest failure has been this last bullet. [*The final bullet on the slide in question reads, "Inventing long-lasting collaborations."*] These workshops are *great* fun, and we all get excited, and we share ideas; we pick up new jargon. I love the—what was it?—"inch pebble", I think I'll keep that one ... probably not very well, though! [*laughter*] We'll have that kind of fun. But the workshop—frankly—is the easiest thing to do. The harder thing to do is to figure out collaborations, or the importation of best practices, across fields, to really implement change. And I hope we don't have too much fun without attending to that long-run problem, because that's really the payoff.

So, thanks a lot—I guess it's lunchtime, Dan?

CORK: It is lunchtime. Actually, if there are one or two questions, we could take those without thoroughly breaking the schedule, but just a couple, and then lunch is in the next room. Yes?

DOYLE: I think that one of the points you made sort of pointed out why I shook my head *this* way and others shook their head *that* way with regard to how stable the instruments are. It really depends on the level at which you look at them. Sure, we've always asked for age, race, sex—we've always asked those questions. But *now* we're asking race with multiple questions, different sets of questions. We make changes at times during the development process that are not opportune. And that's where we get into trouble.

I think we have a sense of an appropriate design-develop-test model that you provided, and we would love to implement it. But, in our lives, the requirements are really not set at the beginning. And not at the level of all the questions, and not at the level of all the flows.

So, just within the instrument piece, we live in a world where we are constrained by reality—the laws change when we're halfway through developing an instrument and suddenly our questions are irrelevant. That's the kind of thing that I can see—from your [Poore's] description, it sounds like that's controlled better when you're developing a product, and know what the product is. But in our case it doesn't seem to be under control. We need to find a way to live within that unpredictability.

GROVES: My third meeting like this one was a wonderful day spent in Palo Alto, overlooking the ocean as I recall, with a set of software engineers for expert systems. And the question on the table was, "Why can't we develop an expert system for questionnaire construction?" That was what went on. It was a great day, we had a lot of fun. The conclusion of the expert systems engineers, at the end of the day, was, "*Perhaps* you people would like to define how to do a questionnaire before you talk to us next time." [*laughter*] And partly that's it. And the immediate reaction is, "Why don't you standardize this stuff? That's stupid; you've been doing this for thirty years, or fifty years. Well, that's your problem, and when you fix that, come back to me." Well, it doesn't quite work like that.

I think the other thing to note is that there's a real distinction in our discussion so far between developing software of generalizable use versus developing an application within a software framework. So some of what Pat was talking about is of the ilk: "What does a user need to know to develop a document in Microsoft Word?" And some of what she was saying was, "What do you need to do to develop Microsoft Word as a software product?" And those are *very* different things. So my friend at CBS News doesn't design Microsoft Word; they do documents in Microsoft Word—to use this metaphor—nightly to get a survey out. So we need to be careful in our discussions on that point.

CORK: This is a good point to stop here, if we can break things off and pick up after lunch.

[*The workshop stopped for a lunch break.*]

UNDERSTANDING THE DOCUMENTATION PROBLEM FOR COMPLEX CENSUS BUREAU COMPUTER ASSISTED QUESTIONNAIRES

Thomas Piazza

CORK: The first block of talks we have this afternoon is going to deal with the documentation problem; later in the afternoon we will turn to the problem of testing. We will have three talks in the general area of documentation, and the first one is going to be by Tom Piazza, who is the head of the Survey Documentation and Analysis project at the University of California at Berkeley. Tom?

PIAZZA: Right; now we're going to talk about documentation. We had a great line at lunch. We said, "What are we talking about?" You know, what are you documenting? And what was our line—"gratuitous complexity," I think it was.

At any rate, that's what we have to deal with. I'm first going to give just a few words of introduction to the instrument documentation problem in general. Then, I'm going to talk about instrument documentation for CASES instruments. And then Jelke Bethlehem is going to be talking about instrument documentation for Blaise instruments, using the TADEQ system.

Incidentally, it's hard sometimes to see this screen—a lot of this stuff that I'm going to talk about you'll find at page 79 in your [agenda] book. At least those first couple of pages give the basic structure of what I'm going to be showing. So you might refer to those if you have some difficulty with the screen.[18]

Now, in terms of instrument documentation in general, I found very interesting this discussion in the morning on testing and documentation for computer programs in general because it is the same basic problem. And we sometimes get caught up in the complexity of these instruments and think that it's completely different. It's not *completely* different, but it's good for us to step back and look at things from a global perspective.

Why is documentation a problem? Notice that it was *introduced* not as "let's talk about instrument documentation" but "let's talk about the instrument documentation *problem*." It is perceived as a problem, and indeed it is. And part of it is the pressure of production work. It does get pushed to the end, and I suppose that's inevitable. But we haven't really dealt adequately with it.

---

[18]As of the time this report went to press, Dr. Piazza's Web site—from which he delivered this talk, and which contains links to additional information on both IDOC and TADEQ—is still accessible at http://sda.berkeley.edu/present/automation. From that page, links are available to the specific pages referenced in these remarks; we have excerpted some of those in this printed report.

And another situation is that documentation really depends on the proprietary tools you happen to be using. Now, many times when you're testing and documenting computer systems and tools ... [that is, in general computer systems testing,] you don't have to use, necessarily, documentation tools that are dependent on a C++ processor, for instance, or something like that. But [in the survey community], because a lot of our CAPI instruments are so tailored to proprietary systems, our documentation is also dependent on that.

And that's one reason I was talking about a possible future—this is an idea that's been floating around for a while—where the documentation language could work off of some common representation of the questionnaire so that you would at least have enough market for developing documentation tools. See, right now, these are all stovepipe-type operations, and it's very difficult for anyone to do documentation without getting to the innards of all this. Now, as we'll see in the TADEQ presentation, they've approached this a little bit in terms of having an XML representation of the Blaise instrument. So they're thinking along these lines also, to some extent. From a computer design perspective, if you're thinking along the lines of, "this is a good model," I just want to make sure you understand we're already thinking about that. It's a possibility, but I would say it's—it hasn't gone anywhere. And I would say that, occasionally, it gets, I would say, "polite yawns." [*laughter*]

On that cheery note ... We're not going to document a pie-in-the-sky model. We have to document the instruments we actually have, the way they run right now. And so let me talk about documenting CASES instruments. I'm going to show you first what some aspects of what an instrument document (IDOC) looks like, and then I'm going to say briefly how one is created, and then maybe a few closing comments before we turn to the next discussion, which will be on documenting Blaise instruments.

If you're going to start with an instrument document, one thing that's useful to do—perhaps you have done—there is an online file that is always available with an instrument document that gives some information on, you know, what it looks like, what indexes are available, it talks about it. And even, down here—it's kind of cute—it even gives you a color-coding that you'll find in there. So, if all of these things aren't immediately intelligible, some of these things you can obtain just by looking at that help file.

Now let's switch to the opening page for our old friend, the SIPP instrument. When you look at the instrument document for the SIPP instrument, this is what it comes up. Here it says the title and so forth, and so forth, and it says, "start over there on the left." And you can get the help file I told you about, which is the way to start. But what I

want to point out first are indexes, because—in a CASES instrument—
it's difficult to document top-down logic because it's a GOTO-structured
language.[19] So about the only thing that can help you here is a variety
of indexes to different parts of the instrument.[20] And it helps you get
some top-down view.

I'll come back in a second to the input screens, but I'll show you,
here, "Items within Modules." Usually, Census Bureau instruments are
designed in a variety of modules. Now, notice the only one here that's
got any text—because Pat Doyle has an interest in that—is the medical
expenses module. So you have various aspects of the SIPP instrument,
and this is one we could look at. And if you go here, you could look at the
various items that are within the module. Notice that the color-coding
comes in here—the red ones are respondent input. In other words, these
are so-called "input items" where the interviewer will enter some data.
And the others have to do with tests and logic and things like that.
And these types of secondary things are what you get when you have
multiple items on a screen—in other words, you have one screen but
many questions on one screen. And that's your clue that that's whats
going in those cases.

So that's what you get for just an overview. Let's look at "alphabetical
order". Well, sometimes that's useful, but there are so many instances.
[If] you have some particular item you're looking for, you can find it.
But [that's] rare, I would say. The best thing are the keywords, where
you can go and find particular items based on keywords. So let's look at
"health insurance" here. And, oh, there they are, a couple of items. But
the problem is that keywords have to come from human intervention.
And human intervention is a rare thing in instrument documentation.
[*laughter*] There, special effort was made to put these up, but that's about
it.

There's an esoteric structural thing here—[the index on] "roster and
section transitions"—that only a specialist would care to look at in terms
of trying to understand how transitions go from one part of an instrument to another, particularly in this type of instrument where we have
rosters. Now, "rosters" mean cycling through people—all the people in
a household. And that adds complexity, and this gives you some indication of how those things go. Again, if you're a specialist, you might find
that useful.

---

[19]GOTO refers to the command contained in some computer programming languages
(the best known example of which is probably the BASIC language) which allows program
execution to immediately jump to another line or another portion of the program.

[20]The indexes available from the SIPP index page are: "Input Screens," "Items Within
Modules," "Alphabetical Order," "Keywords," and "Roster and Section Transitions."

One of the things that I do want to look at here is the index of input screens. What this means is ... let's pick a module here, the medical expenses one again. And the interesting thing here is that this lists only the screens that take input from the keyboard, and this is often what people want to look at—to see what is presented to the interviewers or to the respondents in a self-administered mode. And you can just start here and look at any of them. And what you *can* do, once you get into the guts of this, for every item you have a thing up there—notice it says `ME04` is the next screen in the instrument—is just by clicking this one thing, you can skip through all of the input screens in the instrument. So this is one way to get a quick look about what is there.

And I might add that most of this documentation is really done online—that is, you can see how this hypertext linking takes advantage of this being just a series of HTML files. You can't duplicate that on paper. However, there is an option to produce an ASCII file of documentation. And the default mode is to show *only* these input screens, because that's typically what you have to print out and show somebody because they often want to know what types of questions you're asking. And you might need approval for that type of question. That is a possibility.

But I'm going to switch gears here now, and let's look at an individual item—one of these health insurance items we were looking at [item `ME16`]. [*See Figure II-5.*] And you get a sense of the complexity of what you think might be a relatively simple question. Here we have some of these "fills" that Pat was referring to—in other words, you don't just leave this up to the interviewer. You try to provide the name of the person you're interviewing, or the relationship, and sometimes that affects the question. So you wind up with things like this, [`fill TEMP-NAME`]. What is `TEMPNAME`? Well, you can click on this and go to that part of the instrument, and here you'll notice [it says] "first name last name/you fill," and here at the bottom is an indication that this item has been stored from another item.[21] You see how complicated this gets. So you don't really *know* what's there. But, here, these items tell you where possibly something could have been stored.

Let's go back, and another thing to notice is—look at this here—[`if PCNT le <1>`].[22] Here, the text of the question is conditional. And the issue here is: how much did so-and-so pay for health insurance, or how much did so-and-so pay for health insurance for *you*, or for oneself? And, notice here, [`fill SELF`]—you want to know who "`SELF`" is?

---

[21] To clarify, the "label" on the variable `TEMPNAME` is "first name last name/you fill," and the detail record for the variable points out three preceding points in the instrument where information was stored into the variable `TEMPNAME`.

[22] `le` stands for "less than or equal to" ($\leq$).

<table>
<tr><td colspan="2">ME16   Health Insurance Expenses</td></tr>
<tr><td colspan="2"><u>More detail</u> about this item   <u>ME17</u> is next screen in instrument</td></tr>
</table>

Text of this Question or Item

```
[if PCNT le <1>]
  During the past 12 months, about how much did
  [fill TEMPNAME] pay for health insurance?
[else]
  During the past 12 months, about how much did
  [fill TEMPNAME] pay for health insurance for [fill SELF]
  or others in the household?
[endif]



NOTE TO FR: If someone else in the household pays for the health
insurance that covers this respondent, do NOT try to separate the
amounts for each person.  Just mark N (none) for this respondent
and mark the whole amount when you ask this question for the
person who pays the premium.

ENTER "N" FOR NO PAYMENTS

        @ dollars
```

Description of Instrument Flow
Universe: all adults
    How to Get to This Item
The preceding decision point: <u>ME14</u>
Default preceding item: <u>ME15</u>
    Where You Can Go From This Item

```
Based on the value of this item:
  Go to ME18 if this item = N
  Go to ME18 if this item = 1-99999
Otherwise go to the next item: ME17
```

Figure II-5   Item ME16 from Instrument Document (IDOC) for Wave 6 of the Survey of Income and Program Participation (SIPP).

Well, here we go—what is **SELF**? Depends—it could have been stored in any of these three items.[23] So these are complicated instruments.

Notice here—"more detail about this item." We try to present fairly simple things. But if you want more, well, OK, we'll give you more. [*See Figure II-6.*] And notice the logic associated with it. We see that if the response is: how much do you pay, between \$1 and [\$99999 per year], you go to one item. And if you pay nothing you go to the same place. But look at these "D"s and "R"s—if they don't know or they refuse, then something else happens. Here, let's return to the main description

---

[23] **SELF** would be filled in to read as either "himself" or "herself," depending on information processed at any of three preceding points in the instrument (presumably, the gender questions or recodes).

ME16   Health Insurance Expenses

Return to main description

Commands to Execute After Input

```
<1-99999> [goto ME18]
<N> [missing] [goto ME18]
<D> [missing]
<R> [missing]
```

Context of Item in Instrument

Module name:	medtm.q (Medical Expenses Module)
Section:	LFINTRO
Roster:	persons
Roster level:	1
Preceding item:	ME15
Following item:	ME17

Properties

Item type:	Single input item
Content type:	numeric
Input codes:	1-99999,N,D,R
Missing-data codes:	N,D,R

Location

Roster number:	4
Record/columns:	302/46-50

Keywords

medical expenses; health insurance

Return to main description

Figure II-6   "More Information About This Item" View of Item ME16 from Instrument Document (IDOC) for Wave 6 of the Survey of Income and Program Participation (SIPP).

and see what happens. It goes down here, here's a summary [of] what happens.[24]

Now I'd like to talk a little bit about the logic tools that are available. With a GOTO-based instrument it's very difficult to get a top-down view because, really, you can go anywhere. And, often in these instruments, they do. [*laughter*] So, it's possible with a GOTO language to write a very structured thing. Think of an old BASIC program, with the line numbers—you could go from anywhere to anywhere, the old spaghetti code. CASES is like that.

Why is it like that? Well, it's in response to the development environment out of which it came. In other words, you start with a paper questionnaire, and on the paper questionnaire, you have something where if you answer "yes" there's a little arrow that says "go here" or another arrow that says "no, go *there*." And what CASES did—in the development

---

[24]Specifically, if the respondent answered "don't know" or "refuse" to giving a specific amount paid over the preceding year, they are routed to a question that asks for a (presumably easier to estimate) categorical value, e.g., "less than $500," "$500 to $1000," and so on. After that question is answered, the respondent is routed to the question on direct expenses for medical care (ME18) that they would have gone to if they had given a numeric answer at ME16.

environment of its time—was to make things as close as possible to a paper-and-pencil questionnaire, instead of a computer science approach, which is more what you'll see in Blaise, where you start with a model and a logic. CASES came out in an earlier, a different environment in which we really tried to make it look like a pencil-and-paper questionnaire. And that's also why nobody seemed to think of documentation, because a paper-and-pencil questionnaire is self-documenting, sort of, and the early users of CASES thought that a CASES instrument would be similarly [self-documenting.] And it is more readable than a Blaise instrument, in that sense. But once you allow it to become more and more complex that analogy breaks down, and it has broken down. And that's the problem.

So, at any rate, what do you do with a GOTO-based language? Since you can't really get a top-down view, the most you can do is get a sort of bottom-up view. And what does that mean? Well, a few things. Take a look at this section here—where you can go from this item [referring again to item ME16, Figure II-5]. Now, even in a GOTO-based language, you can always know where you're going *to*—I mean, that's the point. So, at any given moment in the questionnaire you're going to go somewhere else, and that's obvious from the questionnaire itself. And, in this case, notice what we have here—based on the value of this item, you're going to go to ME18 if this item equals "N" for "none", or you're also going to go to the same place if it's between this range; otherwise, you're going to go somewhere else. Now, in this case, you're just going to drop through to this next one, and if they don't know or refuse how much is paid for health insurance you drop down to this one where you try to get them to answer in terms of ranges—ranges of amounts. So, at any rate, from the instrument you can figure out where you're going to go *to*—not a problem.

Trickier is to find out where you came *from*.

Now, here we have the default preceding item—ME15, let's just click on that. OK, bed days, well—this means that that's where we came from. Let's go back, and where does this come from? Now this is not obvious from the instrument itself. But the program that puts together this whole set of HTML files goes through all of the destinations in the instrument. In other words, for every item you know where you go *to*. So, for every item to which you go you maintain a list of all of the other items that said, "go here". Usually, there aren't that many immediate ones; there are many paths to get there, but there are usually just a few items that say, specifically, "go to this item." So you make a list of those. And if there's only one, you wind up here with just one, ME15. In other words, there is no other item in the whole instrument that says, specifically, "go to `ME16`"—the only one is `ME15`, which happens to be right above it. So, in this case, it's simple.

Now what else can you do? One more level of abstraction is provided in the IDOC thing, which is the "preceding decision point." Now, what does that mean? Think of a typical questionnaire, in which you have a series of items—you go from this one to that one, to the next one, to the next one, to the next one. OK? In other words, there's no logical issue here—you just keep dropping down unless, conceivably, you skip out—but that's a different issue. So you just drop right down. So what's useful to identify, sometimes, is: where is the beginning of this series, at which some decision—some logic—could enter? And that's what this particular thing is, the preceding decision point. Now notice here that `ME15` is the preceding item but `ME14` is the preceding decision point. That's the point—let's go back there—at which there is some logic which could affect the flow. And we could look at "more detail" here and see what we have—yeah, sure enough, let's look at some of these things. Here we go ... ooooo, quite a bit![25] ... Here are commands, a set of commands. For the faint-of-heart, we bury this on the second page of every description. But, at any rate, that's what happened.

This is basically all of the logic tools you have available in IDOC, for CASES instruments—as you see, just from the point of view of the bottom up. From the local item—where you can go, where you can come from, and then the beginning of a series for the decision point. Conceivably, you could use—particularly these decision point items—to create a higher-level structure. It would be possible to do that, so that— even in real time, you could say, "I want to map how do I get here, what's involved?" And, from a local point, you would get forward-and-back a little bit.

It would be possible to do that, huh? I don't think I'm going to do it, but ... [*laughter*]

PARTICIPANT: This information is supplied by the CASES instrument versus input by the persons implementing the specifications?

PIAZZA: All of this is really provided by the system itself; we don't really rely on individuals. There is one exception—if you look right above here, where it says "'description of instrument flow universe." This is just an option for people to put in, "all adults." [It's] just an annotation.

Now it would be possible—and there are hooks—for the author to put a lot of this information in as you go. And ideally that's the time to

---

[25]The "more detail" screen for `ME14` includes a complicated sequence of set-up commands that would insert appropriate phrases (e.g., regarding diabetic equipment) into the question, along with instructions for how to handle a key entry of "H," which would bring up a "flashcard" with more detailed descriptions.

put it in, because when the person is *writing* this particular item he or she knows at that point who it's supposed to apply to. A lot of information is available, at that point, and if you have a hook, you put it in, it's carried forward in this instrument document as well. And—if you don't put it in now—no one's going to go back and put it in later. That's the reality.

So a lot of the things we talk about, that Pat talked about, in terms of organizational problems and priorities and so forth—this is one more thing for authors to worry about and do, to put in these bits of information.

I'll give you another, simple example—here, just look at this particular item. "Health insurance expenses"—the long label that we *love* and use in SAS, SPSS, and so forth, a long label for each variable. That's not a part of a [CAPI or CATI] instrument because ... sure, you need a name for the variable, but you don't *need* to put this long label. So nobody puts it in. Now there's a *slot* to put it in, and you could do it, and if you did it would be carried over to the documentation. And it would even be carried over to a set-up for SAS. But, since it's not required to make a [computer-assisted] instrument run, people tend to overlook it.

What we try to do in documenting CASES instruments is—if nothing is specified—we try to manufacture one from the first line of the question. So, here, what would you have? "During the past 12 months about how much did ..." Well, OK—that's not ... I guess you'd get a sense that it's something about how much of something, but it wouldn't tell you too much ... Although I think we're getting more sophisticated [in deriving these default names], so if we just drop this fill [and thus omit the first clause of the question]: "How much did *x* ... "—well, that's not much better. And then you have [other] fills as well, so that makes it more complicated.

So at the time this item is being authored somebody would have a fair shot at coming up with a descriptive label for this item. But one reason people have not done it in the past is that they haven't had any confidence that it would do any good [or] that it would be carried over to the next stage. [There] is a hook in the system to allow doing that, but I don't think it's done very much. Part of the problem is an organizational one. At the Census Bureau—[where] the authors don't ever have to use the data—there's very little incentive to make things easier for people down the pike when they're under immediate pressure to get their *own* job done. So part of it's an organizational thing.

So, at any rate, I think that's about all I have time for in talking about this part. Before I move on, does anybody have any questions about instrument documentation here, that you've seen? Something drastic?

PARTICIPANT: Would you get any major pushback, resistance from the authors, or would it be feasible, to *require* more of the tagging, descriptions, test information into the original authoring?

PIAZZA: Well, that would be for Pat to answer—I don't know—or the authoring staff. Let's see what they would say.

DOYLE: Typically, if the author is providing the label for a particular item in the instrument, the problem is that that's not the person who needs to create it. The person who needs to create it is typically the analyst who invented the question, and it's most efficiently done at the point in time they invented the question. Well, some of our questions were invented forty years ago [*laughter*], and the exercise of going back now and finding someone to go through and provide the nice labels .... It would be *phenomenal* for the user community to have. It's just [that] it's hard to get 'em motivated. You know, this is part of my behavior chain—to get the industry as a whole to habitually include these nice English additions.

The other thing that's user-input here is keywords, which is a *very* useful tool for anybody looking for answers. And if we could do keywords, and feed them into the questions, we could help our users *so* much. And my IASSIST friends are over there, *smiling* away.[26] They know the value of these things; they're phenomenal. But the only person who can really do that well is the person who's designing the question, who *knows* the content and what they're trying to get at. ... It's a real dilemma.

GROVES: When you're doing this, Tom, what are the different users you're thinking of? Who are the different users? And, if you place one user group over another ... who are the readers of this documentation?

PIAZZA: I *wish* I had a good sense of that. I mean, most of them I think are at the Census Bureau. Most of the CASES users don't have instruments that are complicated enough to go through this, I think; many of them just have very straight-ahead instruments, and they still think that they can just read the instrument. So, I'm not aware that many people use it except at the Census Bureau. And, here at the Bureau, I'm not sure *who* uses it. Pat, maybe you could say?

DOYLE: Well, I thought the question was: what was the *intent*? And the intent is quite different than the actual. It has very little usage, at the moment. And this is mostly because very few of the instruments have the additional or optimal labels to facilitate transfer.

Its intent was to serve everybody who needs to use the instrument. Someone like Libbie [Stephenson], who's going to go get the SIPP file

---

[26]IASSIST is the International Association of Social Science Information Service and Technology, and its membership includes social science researchers, information specialists, and methodologits/computing specialists.

for the data archive, the data library, to get it to her users. And she needs to "hand over" the questionnaire. So she could hand over this—either a URL or this document—to a user, and they could sit down at Netscape and figure out what's in the survey. The intent is also to have something to hand over to [the Census Bureau's Center for Survey Methods Research (CSMR)], to the cognitive people, to see the different variations on the questions, and assess whether or not there are problems ... so, to learn what's in there before they begin to develop alternatives. It's basically for anybody and everybody who would have normally wanted to look at the paper instrument.

PIAZZA: There's an additional problem, too, and that is—especially in terms of federal data sets—all of these variables are derived, or censored, or recoded, or something, so that the data set you get is far removed from the original questionnaire. So, one thing that we are working on in terms of the documentation of the data sets is to allow links back to the document for the questionnaire, which is related but not the same thing. So that becomes a [more] complex project. But that can be done as well.

BANKS: It begins to sound as though one wants to have many, several, different, independent kinds of documentation.

PIAZZA: Yes.

BANKS: One documentation for the analysts, one documentation for the potential user, and so forth. And that these shouldn't necessarily interact at all.

PIAZZA: That's true, that's true. In terms of the analyst, that's focused on the data set, huh? So, occasionally, you may want to go back and say, "Let me see where this variable came from." And you click, and you could go back to the original instrument. And in terms of many of the users on the production side ... well, that's another issue. The developers may want *all* of this information, particularly focusing on some of the logic issues.

One of the by-products of creating the instrument document is that you have this whole list of—this matrix of where you can go, from where *to* where. And if it turns out that one of your items is not the destination of *anything*, this becomes rather obvious. And in fact the process prints out a warning message, so that you'll know that nobody can get to here. So that's one diagnostic aid that's a by-product of doing this kind of documentation.

Ideally, too, people should be able to pick the level of complexity they want to look at. We sort of isolated this by having the main things on one page; then you go to another page that's more details. But, you know, ideally, you could also say let me *click-click-click* specify directly what I want to look at—just *those* aspects of each instrument. So, all of these things are possible, but result in a more complex interface.

PARTICIPANT: It sounds like a lot of what you've been talking about is to get people to use, to put these sorts of hooks into the programs. And I think this gets back to what Jesse [Poore] was saying this morning, about the specifications documents, getting the questions together and putting labels to put with questions.

I haven't heard anybody say anything about this today but there's an effort at the Census Bureau to put together a corporate metadata repository, which would be a place where you could put those specifications. And it drives part of the software development process—be it CASES or Blaise—to get the labels onto data at that point of the process, to make it easier to either analyze data or to document it on the Web . . . [*trails off to inaudible*]

PIAZZA: Is it? I don't know. I mean, there is a movement to use . . .

DOYLE: Well, the corporate metadata repository isn't focused on the needs of household surveys as much as it is business surveys, so it doesn't have all the components. But the basic idea is right—we are using other tools for the same purpose, particularly in database access tools.

[You] had said something earlier about using this to drive development? Well, part of the design of the database is to include these additional pieces of information that are only used for documentation purposes—the labels, keywords, and that sort of thing. But, again, just having the place to put it doesn't necessarily mean that it's going to get put there. So we still have to . . . we can do everything we can do to save it and preserve it once it's put there. We just need to get someone to put it there.

PARTICIPANT: Right. But the only way you're going to get people to put it there is to make your software system . . .

DOYLE: Well, if you try to force someone to put it in—in other words, if you . . .

PARTICIPANT: . . . I guess, a penalty for them *not* to put something in.

DOYLE: OK. But what they could do is decide not to use the tool, and go do it in their word processor. They'll find a way around it, if they don't want to do it.

PIAZZA: Given that it's more work, right.

DOYLE: And they'll do that if they don't see the value. So we need to teach the value of doing this; all these individuals need to see what can benefit them, and the use of their data, by their taking the extra step forward.

PIAZZA: Right. OK, I'm about out of time here, so what I want to do here is to make one mention here—that you should be aware of the Data Documentation Initiative (DDI). Instrument documentation is somewhat different from documenting datasets, but there are a lot of things in common, and these two things are aware of one another. So, that's one thing to keep in mind, is there are some standards for documentation, and we're trying to make use of those.

Then, the other thing is that it is possible to do some sort of convergence of documentation systems. I won't go back to the original flowchart. But some of the stuff that we do is related to what you're going to hear about now, with Blaise. For instance, what a CASES instrument does ... there's a program that creates an ASCII file that's got all of the specs in it, which is then put together. And then—what you're going to hear about in TADEQ—that also generates an XML representation of a Blaise instrument. [So, ideally, this migration to a common intermediate format might make it possible to develop more universally applicable documentation tools.] And maybe more would get done because, remember, this is a very small market. And unless you have a broader range of applications, a lot of these tools just don't get developed.

So I think I'll stop there and later on I can take any questions you might want.

POORE: I keep seeing a graph, and each instance being a walk on the graph.

PIAZZA: A what?

POORE: A walk on the graph. Why isn't that enough documentation?

PIAZZA: You mean, a flowchart sort of thing, or what?

POORE: A directed graph, from beginning to end of the instrument. And each instance of the survey, you can peel off of the graph.

PIAZZA: That's true, but there are so many graphs.

POORE: No—one graph; many walks.

PIAZZA: OK, but so many walks! I mean, that's our problem.

POORE: One per interview ...

PIAZZA: But if you have 100,000 interviews, that's hard, you know?

POORE: Well, OK; I just, it doesn't seem that hard to me.

PIAZZA: It's doable, but who's going to look at it? I'd love to talk about these things but we have to move on, so let me close this down.

PARTICIPANT: But maybe a lot of the interviews follow the same path?

PIAZZA: Oh sure, oh sure.

PARTICIPANT: And that knowledge would narrow things down a lot.

CORK: Let's have one more [question], and then we have to make a switchover in technology here.

PIAZZA: Right ... Why don't you start now, and I'll just take any final questions here?

BANKS: Going to Jesse's point about the many possible ways through the graph, it would seem that one would probably go a long way down the road towards documenting things adequately if you had ... if you treated each module as a vertex in the graph, and then the paths linking modules in the graph, going from one to the other. One has one kind of documentation for a flow, and I think you would want to have pairwise documentation for pairs of modules. And I don't know if you need to go through the exercise and effort of documenting all possible paths when getting connections between modules of questions could carry you a very long way down the road.

PIAZZA: Right, right. And we have to find some way, like that, that will give a close enough approximation. But partly it's just to understand the major, just the overall view of the instrument. It's not the case that we need a full flowchart of everything because that gets very complicated. And a lot depends on the design of the instrument. If it's really a modular design, then of course it's a lot easier because you just go from one to the other. Then, within those, recursively, you have the same setup.

DOYLE: The other part of the issue, of using just a flowchart, is that it's critical to fill out the words. And once you try to fit all of the words in a flowchart, you've taken up a lot of the visual real estate with the words, and you can't get a lot of your paths in there. So what you wind up lots of times is a visual of the flow and then the text of the questions. Which, if you've got these in some nice coordinated software package ...

PIAZZA: In synch, right ...

DOYLE: In synch, then it's probably all right. But that's where it needs to go. but it just can't be a simple flowchart. Because there's too many words to put into one box.

PIAZZA: Actually, TADEQ does some of that, where you go through and then click on the side; you can select the view and get the text of the question. But, as you see in the examples I've presented, sometimes the text of the question—as in the instrument—isn't all that aformative because it's conditional text or it has all of these fills. So it's really hard to know, for a particular interview, what really was done.

PARTICIPANT: [*inaudible*]

PIAZZA: Maybe your point is: I guess we don't have to, if it worked, if they got to the end all right. It's true, the amount of energy it takes for documentation. It makes you wonder, huh?

GROVES: The one thing about what we do is that we are utterly uninterested in a single interview.

PIAZZA: Yes, yes, usually.

GROVES: What we're interested in is the information that is taken away by summarizing. So in a way I think Jesse's point is well-taken; if you had a visual technique that could readily identify where 90 percent of the cases are going to appear, or 80 percent of the cases, or the modal case—that would be very valuable. I don't know how you would do that in one visual appeal, given the nature of these trees.

PIAZZA: The one case where we really are interested in the individual record is to make sure that the logic was followed for that individual records—in other words, that the averages are based on appropriate answers to the things that we're answering.

PARTICIPANT: It's of interest to the users, I suppose.

PIAZZA: Could be.

PARTICIPANT: One other role for individual traversals of the graph would be for what's called regression testing. So that if you could save these traversals of the graph, for this information, and then you redesign some features, these are scenarios that you could then run through to make sure that what you're doing satisfies, meets what's previously been done.

PIAZZA: OK, all set? Thank you.

THE TADEQ PROJECT: DOCUMENTATION OF ELECTRONIC QUESTIONNAIRES

Jelke Bethlehem

CORK: OK, now that I've learned the difference between "Shut Down" and "Restart" in Dutch ... [*laughter*], we're just about set.

Our next speaker is going to be Jelke Bethlehem, who is a senior advisor to the Department of Statistical Methods at Statistics Netherlands, and is past head of the Statistical Informatics branch there.

BETHLEHEM: Well, let me start by thanking the organizers for inviting me to participate in this conference. Up until now, it's been quite interesting and a lot of fun to be here for an exchange of interesting opinions.

What I'm going to do is ... well, I think Tom Piazza almost gave my presentation, so I don't have to add much more. [*laughter*]

What I want to tell you is a little bit about some research we did in Europe with respect to questionnaire documentation and—to be more specific—questionnaire instrument documentation. The background of this is the same as was already discussed this morning.

We changed, in the Netherlands and in Europe, to computer-assisted interviewing—and in particular to computer-assisted personal interviewing with laptops—in the 80s of the previous century. And we had, from [that] moment, a growing complexity of questionnaires [and] growing size of questionnaires. We have seen, a famous example, instruments with 10,000s of questions. And indeed the old paper questionnaire was more-or-less self-documenting. But these large, complex questionnaires have become more and more of a problem from the point of view of documentation.

So the basic question is: how can you make, for a complex, large questionnaire instrument, a human-readable documentation? And we wanted to have a solution for that and started a project to develop a prototype for that, and that became TADEQ. So we want to generate, to create a software tool for questionnaire documentation. And, when we thought about it, we realized that once you have a tool for questionnaire documentation it's also a very useful tool in the process of developing a questionnaire. Because more or less the same aspects with respect to documentation, and getting insights into complex structure of the questionnaire, are involved in both the final documentation and in the process of developing it.

What we tried to do—as Tom also explained—was that we started in a later phase of thinking on this issue, and tried to have a more-or-less object-oriented approach in developing a questionnaire documentation tool. We see a questionnaire instrument as a collection of objects—all kinds of objects you can have in a questionnaire. Questions, checks, route instruction, computations—whatever you can think of. And all these objects are part of a kind of questionnaire execution tree.

So here I've named a number of these types of objects. You have questions—the various types of questions [such as] open, floats, numeric, etc. Route instructions, which can either be GOTO-oriented instructions as in CASES or IF-THEN-ELSE structures like in Blaise. We have checks in the questionnaire structure; these were not mentioned too much this morning, I think, but these are a valuable, extra advantage as compared to a paper questionnaire. You are able to detect inconsistencies while you are carrying out an interview, and are also able to correct incomplete answers in the course of an interview. If you have to do that later on, in the office, it's almost impossible to do that. So we feel that, with respect to checks, a questionnaire instrument can add to the quality of the final collected data. You have computations, you have loops. And

```
1. Are you male or female?
   Male . . . . . . . . . . . . . . . . . . . . 1    Go to question 3
   Female . . . . . . . . . . . . . . . . . . 2

2. How many children have you had?          - - children

3. What is your age?                          _ _ years

Interviewer: If younger than 17 then go to question 7

4. What is your marital status?
   Never been married . . . . . . . . . . . . 1    Go to question 6
   Married . . . . . . . . . . . . . . . . . 2
   Separated . . . . . . . . . . . . . . . . 3
   Divorced . . . . . . . . . . . . . . . . . 4    Go to question 6
   Widowed . . . . . . . . . . . . . . . . . 5    Go to question 6

5. What is your spouse's age?                 _ _ years

6. Are you working for pay or profit?
   Yes . . . . . . . . . . . . . . . . . . . . 1
   No . . . . . . . . . . . . . . . . . . . . 2

7. Do you regularly listen to the radio?
   Yes . . . . . . . . . . . . . . . . . . . . 1
   No . . . . . . . . . . . . . . . . . . . . 2

END OF QUESTIONNAIRE
```

Figure II-7 Portion of a sample questionnaire, as it might be represented on paper.

we apply, in Blaise, a more-or-less modular structure, in that we have one big questionnaire but can divide that up into sub-questionnaires, and those sub-questionnaires can be divided, so that you have a more modular approach that I think is very important when you're working on large questionnaires.

Just a simple example—maybe not very well readable in the back of the room. But this is the typical paper questionnaire as it existed a number of years ago. [*See Figure II-7.*] You have the various types of questions, and you have routing instructions that can take two forms. You have these jumps attached to possible answers or you have, say, interviewer instructions that tell the interviewer to go elsewhere in the questionnaire if some condition is satisfied.

Well, this type of questionnaire—if you programmed it in CASES you would have something like that. [*See Figure II-8.*] Of course, you will find the definition of your questions, and the routing instructions,

```
>Sex<
    Are you male or female?
    <1> Male                    [goto Age]
    <2> Female
    @

>Children<
    How many children have you had?
    <0-10>
    @

>Age<
    What is your age?
    <0-120>
    @
[@][if Age lt <17> goto Radio]

>MarStat<
    What is your marital status?
    <1> Never been married    [goto Work]
    <2> Married
    <3> Separated
    <4> Divorced              [goto Work]
    <5> Widowed               [goto Work]
    @

>Spouse<
    What is your spouse's age?
    <0-120>
    @

>Work<
    Are you working for pay or profit?
    <1> Yes
    <2> No
    @

>Radio<
    Do you regularly listen to the radio?
    <1> Yes
    <2> No
    @
```

Figure II-8   Portion of a sample questionnaire, as it might be represented in CASES.

the GOTO-type instructions, in your code, either as separate instructions that instruct the system to go elsewhere or jumps connected to the outcomes of questions.

If you would do that in Blaise, you would get a bit different piece of code. [*See Figure II-9.*] In Blaise, we [felt] that it was important to separate the definition of the questions from the logic of the questionnaire. So we have to separate two parts of the questionnaire: we have the field section in which the questions are defined and every question has a name and unique identification. And then we have the rules section, which tells the system what to do with all these questions. You can ask a question, and you have route instructions like, "if the sex is female then ask the question about children." We also have checks in this rules part, like if marital ... no, actually there aren't any checks in this particular piece of questionnaire. But there are different types of routing instructions here that allow you, in a more structured way, to indicate what question has to be answered under which condition.

Now you see that routing instructions can be very different in CASES and in Blaise. But there are also other systems in the world, and we have even seen hybrid systems that support both GOTO-type and IF-THEN-ELSE-like instructions, which makes things even more complicated, at least if you are looking at that from the routing logic point of view.

But, in the end, it's true that the routing structure is a graph, as was already mentioned just a moment ago, and the vertices are the various types of questionnaire objects you have, and the edges are the possible transitions—possible moves—from one part of the questionnaire to the other. It's not a general type of graph, a questionnaire routing graph ... there are some limitations. It's an acyclic graph; it is not—at least I think it should not—be possible to be able to really jump back in your questionnaire, and making loops in your questionnaire. That would lead to a very long interview, I think. It's a directed graph, so the flow is from the top to the bottom; you never go up again, it's always going down. There's one start vertex, the beginning of the questionnaire, but there are multiple ways out; you can exit the questionnaire at many points, depending on whether conditions are satisfied. It's a connected graph; each point can be reached from the beginning—I hope—and each path is a possible route.

Well, I said, "I hope"—I think this is one of the problems that comes up when complex questionnaires are designed, that it is not always completely clear whether each point can be reached from the beginning of the graph. The logic can be so complex, and conditions determining the routing through the graph may contradict each other. So I think it would be very important to have [a tool] in the analysis, in the development stage of a questionnaire that would allow you to detect whether there are

```
DATAMODEL Example FIELDS
  Sex      "Are you male or female?": (Male, Female)
  Birth    "How many children have you had?": 0..10
  Age      "What is your age?: 0..120
  MarStat "What is your marital status?":
           (NeverMar "Never been married",
            Married   "Married",
            Separate  "Separated",
            Divorced  "Divorced",
            Widowed   "Widowed")
  Spouse   "What is your spouse's age?": 0..120
  Work     "Are you working for pay or profit?": (Yes, No)
  Radio    "Do you regularly listen to the radio?": (Yes, No)

RULES
  Sex
  IF Sex = Female THEN
     Children
  ENDIF
  Age
  IF Age > 16 THEN
     MarStat
     IF MarStat = Married) OR (MarStat = Separate) THEN
        Spouse
     ENDIF
     Work
  ENDIF
  Radio

ENDMODEL
```

Figure II-9   Portion of a sample questionnaire, as it might be represented in Blaise.

[any] points that you can not reach. It would help you greatly, I think, simplify your questionnaires by getting rid of those things. Or, maybe, it would point out an error in the graph that you could fix.

What would such a graph look like? Well, this is a simple example of a small part of our Labour Force Survey. [*See Figure II-10.*] And it shows you also that all these different paths can have great variation in length. For example, here, if you are lucky, by choosing the "right" answer [at node 5 or 9] you can skip a lot of questions and you can go to the end immediately. And if you are so unlucky that you go *this* way, then you have to answer a lot of questions. But it helps also, of course, in organizing the fieldwork; people would like to have some impression of how long taking the questionnaire interview could take. And then, here, you could see that it could be very short or very long.

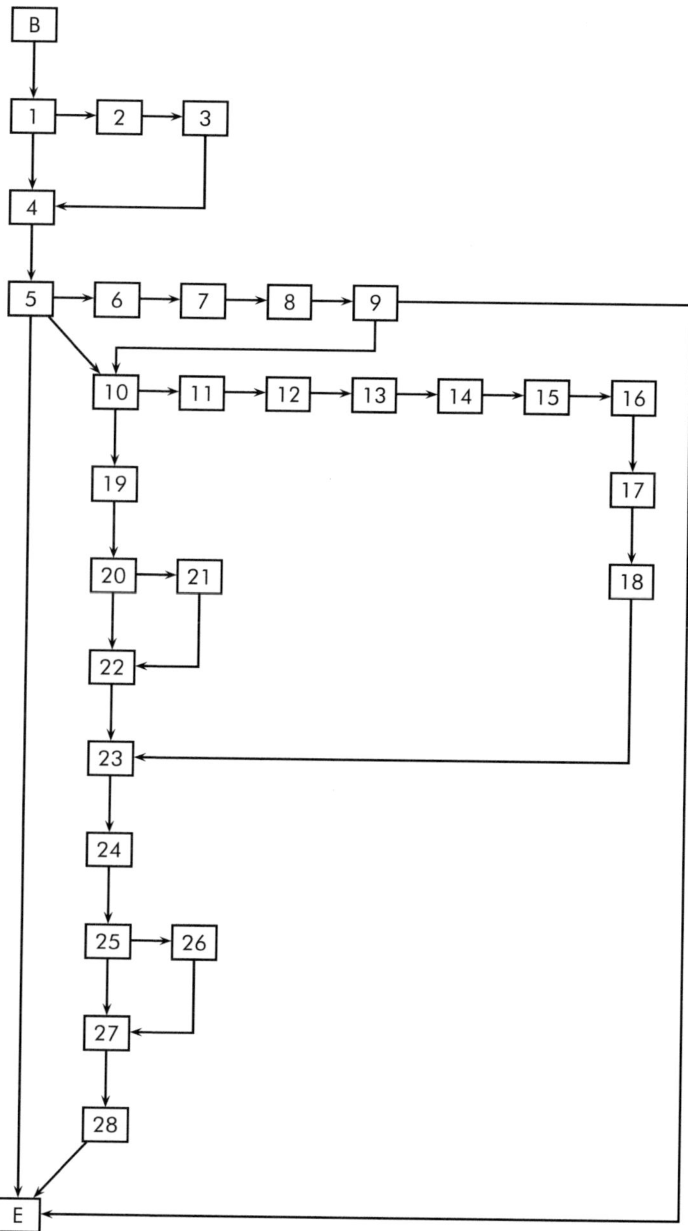

Figure II-10 Hypothetical routing graph of a questionnaire.

SOURCE: Workshop presentation by Jelke Bethlehem.

But this is only limited information; it gives you a global overview of the graph. Of course, you would like to see a little more than this; then, you could think about making a flowchart. Well, everybody knows what a flowchart looks like; the green boxes indicate questions, and these diamonds indicate decision points where you can go to different parts of the questionnaire depending on the condition, whether or not it is satisfied. This can be very illustrative but, on the other hand, also here, information is limited. You can only see the name of the question, and in a very short way the condition that should be satisfied. There is not much more room here to add more information there.

These are some thoughts that we had at the beginning: what shall we do? Do we want textual documentation, like Tom showed? Or do we need graphical information? What do we need? This was the reason this TADEQ project was set up. It was, of course, a European project, set up and funded in part by the European Union.[27] Some of the objectives were that—of course—when you make a tool, it should work in such a way that documentation is automatically generated. You should try to get into a situation where people don't have to do a lot by hand; it takes too long, is costly, is error-prone, etc., etc., etc. You try to design the package—the tool—in such a way so that it would be open for many different systems. So we tried to design an open tool, that is not meant for Blaise only but for any package that can support the language we designed for this tool. Also, we realized that there are different types of documentation necessary, probably, for different types of users—and there are lots of types of users—who need different kinds of information. We had the impression that the users might need both textual and graphic information—that one is not enough and that you need the other. And we were wondering what the users really want: documentation on paper or electronic documentation? Is it sufficient—we hope, at least—to just have electronic documentation, to give them the graphs [and] the hypertext documents, and would that be sufficient?

In order to find out what the users really wanted, we set up this project. And let me just mention the partners in this project: Statistics Netherlands, Office of National Statistics in London, Statistics Finland, National Statistics Institute of Portugal. And another thing about this project was that we had a university in it which was very experienced in computer science, in graph drawing algorithms—something we were not very experienced in. But [it was] nice to work together with someone

---

[27]The TADEQ project involved a consortium of statistical organizations in five European countries: Statistics Netherlands, the Office for National Statistics (United Kingdom), Statistics Finland, the Instituto Nacional de Estatistica (Portugal), and the Max Planck Institute (Germany). [The partners from the Max Planck Institute later relocated to the University of Vienna (Austria).]

from a totally different type of activity; that helped us to do this work. And, indeed, when we contacted these people, they shouted at us that they had already done a lot of work with this that would be very useful for us, but they didn't know. [They argued] that they had a problem but not an application, and we had an application but not the theory behind it. And they have already found out how to do it. So, it was a nice marriage, you could say.

But what we tried to do in this project was find out what the users really needed. We had a look at some of the CAI packages; we tried to design some sort of neutral language for specifying what's going on in the questionnaire instrument. And, in the first prototype, we implemented some documentation functions and, in the second prototype, we introduced some analysis functions. What we found out in our user requirements survey—a survey in which about 100 users of CAI systems were involved—it was clear that different users required different types of documentation. There are a bunch of different types of users—the developers, the managers, the supervisors, the interviewers themselves in the field—[and] all need documentation. And the analysts using the data later on. And they made clear to us that they need both paper and electronic documentation. For example, an interviewer in the field with the laptop, they say that they want documentation on paper because they can not do the interview on the laptop and consult the electronic documentation at the same time. So they want something on paper that they can use if they need it.

There was a need for detailed information about the questions—well, that's obvious, of course. But it was also was important that, for every question, it would be clear what the answer universe was. That's how we defined it: what kind of people are answering this question? What is the sub-group of people, [and] what are the conditions leading to this question? And there was a need for detailed information about the routing structure, and there was—not that heavily expressed—but some need for tools to analyze the routing structure.

OK, we looked at some different CAI software but the results were not too dramatic, I think. Mainly we see two types of packages—the IF-THEN-ELSE-oriented packages and the GOTO-oriented packages—and some with the hybrid approach using both. Blaise is IF-THEN-ELSE, CASES GOTO-oriented. We wanted to look at some other packages, but they suddenly disappeared from the market, so ... it was hard to have a look at them.

```
<qobject number="3" name="Children">
  <question>
    <text><![CDATA[How many children have you had?]]></text>
      <numeric>
        <range lower="0" upper="10"/>
        <integer/>
      </numeric>
    <status>ask</status>
  </question>
</qobject>
```

Figure II-11 Sample question, as coded in the Questionnaire Definition Language used in the TADEQ project.

And then we decided, after looking at some of these packages to design this XML-based specification of the questionnaire instrument.[28] Why XML? Well, I don't know; I think it's an obvious development. At the moment, if you look around, wherever you want to do something with metadata—[to] say something about data—XML is an obvious instrument to do that. I mentioned some other initiatives that also have the idea to do something with XML to specify the metadata structure. So I'll skip over some of those things.

To give you a quick overview—how is a questionnaire defined in QDL, the Questionnaire Definition Language? [*See Figure II-11.*] There are questionnaire objects, and in this case, it's a question; it has a text. It could have many texts in different languages. It's a numeric question, so it has a lower bound and an upper bound. It's an integer answer, etc., etc. And you could have special items, like this question is asked, or it's computed, or so forth. This is a limited example.

This is how the routing structure is defined in the Questionnaire Definition Language. [*See Figure II-12.*] You have a questionnaire object called **split**; it has a condition, sex is female. So the jump is only carried out if the sex is female. And if this condition is true, this part of the questionnaire is carried out, and if it's not, the **if_false** part is carried out. There is no **if_false** part specified here so it's skipped in this case, just to keep it simple.

You also see, generated, is a list of all questions involved in these conditions, which may help in the analysis—which questions determine the routing in the questionnaire, [and] which are involved in the checking of the answers? These are the more important questions. It's all based on a Data Type Definition (DTD) in XML, which in fact turned out to

---

[28]XML, the Extensible Markup Language, is a successor to the more familiar Hypertext Markup Language (HTML) in which most pages on the World Wide Web are coded.

```
<qobject number="2" name="Condition">
  <split>
    <condition><![CDATA[Sex = Female]]></condition>
    <if_true>

      <qobject number="3" name="Children">
        <question>
          <text><![CDATA[How many children have you had?]]></text>
          <numeric>
            <range lower="0" upper="10"/>
            <integer/>
          </numeric>
          <status>ask</status>
        </question>
      </qobject>

    </if_true>
    <involved><![CDATA[Sex]]></involved>
  </split>
</qobject>
```

Figure II-12 Sample route instruction, as coded in the Questionnaire Definition Language (QDL) used in the TADEQ project.

be very simple—one page, just to give you a quick look. A questionnaire consists of specification for a number of languages, descriptions, some attributes, definition of local variables, and then one or more questionnaire objects.

And what's a questionnaire object? A questionnaire object can be a question, split, computation, check, loop, statement, GOTO, or a subquestionnaire. And what's a question? Well, a question can be closed, numeric, open-end, various types of questions. And below here you see that a closed question consists of a number of items, and every item has a code and a text, etc., etc. So, this is a way you can quickly define the structure.

So this was the starting point, for the input to the TADEQ tool. Once you have your questionnaire specification in the XML structure, TADEQ can read it and it then has three elements. It has a structure tree; that gives you a schematic overview of the questionnaire. You can use it to navigate through your questionnaire—to open branches, to close branches, to open sub-questionnaires, to close sub-questionnaires. You can select parts of this structure tree, and then for these selected parts you can have a look on the question documentation that's really focused on the question definitions. It's a textual documentation. We also specify the universe for each question. So there's unlimited information on the route of a questionnaire.

On the other hand, you can choose the graphical documentation on selected parts by choosing another option, and then the focus is more on the routing structure and less on all the details of the questionnaire.

So this is more or less the same diagram as Tom Piazza showed; this is what we intend to do. The TADEQ tool as built is completely separate from the Blaise system. What we have now is that the Blaise system generates the XML file that is then read into TADEQ. And any system that can generate an XML file—in the proper data definition structure—can be input into TADEQ. And then TADEQ produces a tree view, and you can either go for text or the graphical overview to have your documentation.

This is a quick look at the structure tree. [*See Figure II-13.*] It's like the tree you have when you're looking at Windows Explorer; it's a familiar metaphor, I think, for when you're looking at a questionnaire. This is the questionnaire in a kind of collapsed state, you could say.[29] Blaise is a modular language, which has questionnaires, sub-questionnaires, sub-sub-questionnaires. And these squares denote sub-questionnaires. So what you see here is the questionnaire at the main level. And I think these are important things if you develop large questionnaires—to get a global overview of the questionnaire. So "person" here doesn't mean a single question named "person," but a whole block of questions about the person. There's a whole block about work conditions; there's a block about school; there's a block about travel, etc., etc. This is the main view of the questionnaire, and by simply clicking on the icons, you can unfold these three—maybe not all of it at the same time—but you can focus in on the block you want to study or take a look at. This is the view completely unfolded, and you see all single questions here; you see text, computations, checks, all sorts of things. And the routing structures connect them, saying which question you have to go to.

This is mainly the part of the system you use to navigate, to go to a certain part of the questionnaire. But, at least, we've found out that many people like this part of it; you can print this, and it gives you a good overview of the questionnaire.

If you generate textual documentation from this tree, it's [in] HTML format, generated as an HTML format for a selected part of the questionnaire. It has the text of the question [and] some status information; for the closed questions, it mentions all of the possible answers. And also—but not on this screen—it would also show special answers like

---

[29] For sake of brevity, the screen shot in Figure II-13 jumps ahead to the fully-unfolded questionnaire Dr. Bethlehem refers to later in the paragraph.

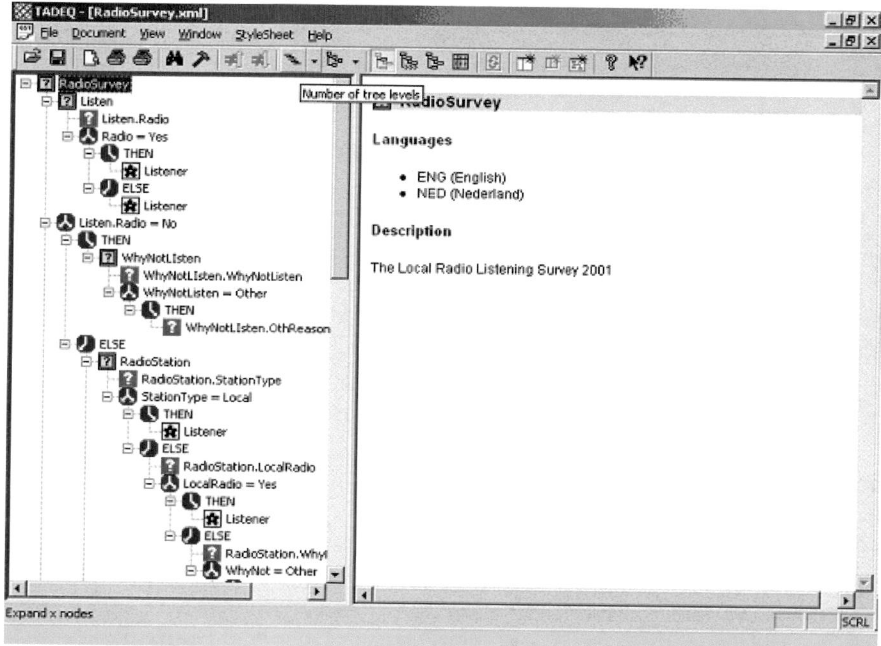

Figure II-13   Screen shot of TADEQ applied to a sample questionnaire, with some sub-questionnaires unfolded.

"don't know" or "refuse." And you also see the answer universe here; this is a very simple one, the condition is age greater than 12. So this question will only be asked of persons 13 years or older. It's an HTML file, and the layout is controlled by cascading style sheets, so if you want a different layout you can just create your own format.

This is part of the graphical documentation. Here, again, the main overview corresponds to the main tree overview. So the green blocks are the sub-questionnaires, and the blue symbols indicate decision points in the routing structure. It's interactive; you can click on the blocks and they will unfold. And this is what you'll get if you unfold the complete questionnaire for this example. You can, of course, zoom in and out if you can't read it.

And there's one other thing: that this information you have can be displayed in two modes. One mode is what we call the "techno" mode; [in that case,] you get the mathematical expressions, the question names. So it's fairly compact but for the designer of the instrument maybe more illustrative. Or you can choose the "textual" mode, and then you get the complete text of the questions in these blocks. And there's more textual information there, which may be easier to read for those people who are not really into designing the instrument but just want to know what this question is all about.

You can see that the amount of space you have is limited, so you can not do everything. But the system has the option that allows you to open an extra window on the screen, and when you click on it, then you get all the information about one specific object, if you want it.

And now the analysis elements—well, "analysis" may be a big word for this but it turns out to be quite useful. This is just a frequency distribution of the types of questions you have, the number of checks, the number of computations, number of splitting points, number of sub-questionnaires. There's a frequency distribution.

This may be a little bit more important. This gives you information with respect to questions and in which types of checks or computations they are involved. So, for example, the question marital status is involved in one check, so it's used in one check, so it might be a more important question than some other questions. And, also, the distance to work is involved in a check. Question age is involved in three split points, so if question age is answered wrong, then you know that things could go wrong elsewhere in the questionnaire.

This is what we were talking about a while ago; this is the length of the path through the questionnaire. [*See Figure II-14 for a sample of the route statistics.*] There is usually a large number of different paths through the questionnaire. And when we use this tool for real-time, real-life surveys, then usually the total number of possible paths is so big that formatting it in a normal programming language is impossible. It's amazing, if you look at the large numbers that appear there, for maybe even moderate questionnaires.

How many different paths are possible through a questionnaire? Well, this is a very simple questionnaire; there are paths with four questions, there are paths with thirteen questions. And the median is somewhere here, nine questions, and there are 74 paths with length nine questions.[30] If you see these large numbers sometimes, you think, "Is this really what we wanted to do when we designed this survey questionnaire? And how about the poor statistician who has to analyze the data later on?" [*laughter*]

OK, there are some things that are still in the process of being developed, that are not yet ready. We can count the number of paths, we

---

[30]The questionnaire being described here—that Dr. Bethlehem analyzed directly using TADEQ during the presentation—is different than the one for which the screen shots in Figures II-13 and II-14 are available.

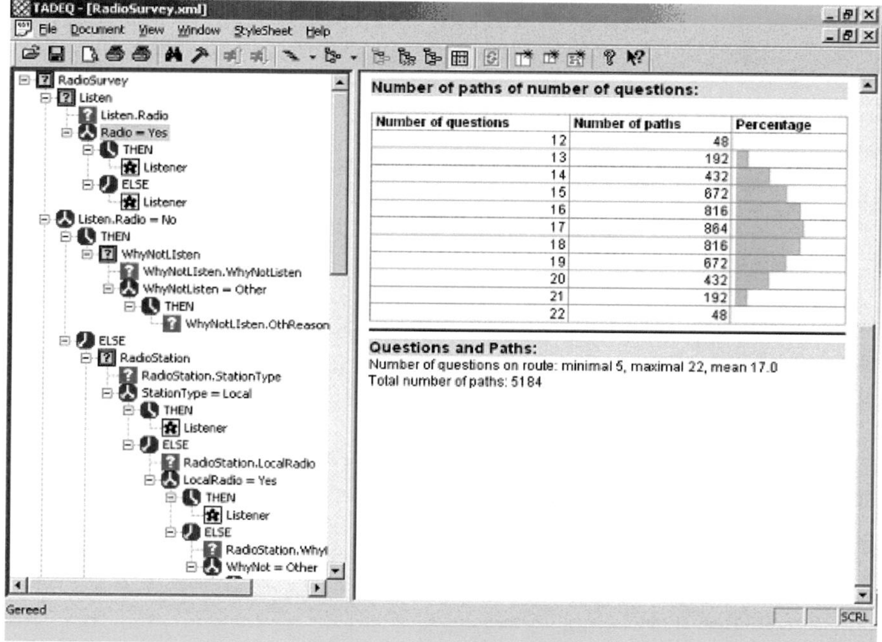

Figure II-14   Screen shot of some route statistics generated by TADEQ for a sample questionnaire.

can compute the length of the paths. And what we also want to do is to do a weighted computation—assign weights to each questionnaire object and then do the weighted distribution of the lengths of the paths. This is important because if you have some kind of simple model capable of assigning to every object how much time it takes to execute—to ask the question—then you would have a distribution of the possible length of the questionnaire interviews. That would be useful in the field; that is what people always ask for.

Because we have the questionnaire instrument now in a convenient format, the XML tree, it also has become possible to compare various questionnaire instruments. That was not really the objective of the project but came out as a sort of side-product, but it is a very useful side-product. So you can feed two questionnaires to the tool, you can compare them, and it will try to find the inconsistencies between the two. And it indicates inconsistencies between the two instruments with different colors. For example, here's a check on marital status; it's red, and that means it's not here. There's a green one here, and it means it's not *there*. And the yellow one—distance to school—is in both, but the texts, the contents are different, one word different from there. This is a very easy tool to compare, for example, instruments for two consecutive years, or something, and you want to know the changes made in the two instruments very quickly.

So, some of the conclusions ... how much time do I have, Dan?

CORK: Keeping with equal time, you have about fifteen minutes.

BETHLEHEM: Fifteen? OK, so I'll continue.

We think that this type of tool could fulfill a need for electronic questionnaire documentation. I also have to stress that this is now a prototype; it has to be developed further. But when we are playing with it, we have the feeling that it can help. And we also feel that we have to do something because hand-made documentation is *out* of the question. I have seen examples of that—thick books for the Labour Force Survey, hundreds of pages of documentation that took three months to prepare. And then we had the next version of the instrument and it was out of date. And also a lot of errors in it.

Because we have something like this we think that it will help promote standardization of questionnaire documentation, through your organization and even between organizations. It could even—as Tom [Piazza] also indicated—get to different output, documentation output for different systems. And, of course, we have the feeling that the analysis tools improve the quality of the data collection instrument because it helps you to find things wrong when you are designing it. But, not all problems are solved yet in this prototype.

To mention some more things ... If you have a GOTO-oriented questionnaire and you want to compute the distribution of the path lengths, you might be in for some trouble, I think. It is a very complicated thing to do if you can jump back and forth through a questionnaire, and even more so if you have the combination of IF-THEN-ELSE and GOTO, where you can use GOTO to jump in and out of IF-THEN-ELSE branches. It's, I think, impossible and I think people should forbid it. [*laughter*]

Weighted analysis of path lengths can also be a very time-consuming task if you have to go through all possible combinations and do a weighted calculation. There should be ways to do something about it; we are now experimenting with sampling the routing graph, and we have the feeling that that might help in some way to get more or less the same results.

Detection of impossible routes could be very important in the design phase of the questionnaire. At this stage, we're not capable of doing this in the TADEQ prototype, because what does that mean? It means that you have to go through the graph and—everywhere you encounter a logical condition [or] expression—you have to evaluate it and go to the next condition, and compare the results from this next condition with the previous conditions. And you get a stack of conditions, and [when] you want to know more about that, you have to write some kind of expression parser. And [at that point] you are almost rebuilding your computer-assisted interviewing system. And we didn't want to do that. And moreover these conditions can be very complicated and could even depend on things not in your questionnaire—for example, on the data set of another questionnaire, or a random number generator, or whatever. So you don't have complete control over all these expressions. I think it could be done, to some extent, but you would have to think about the best way to do it.

What we'd like to do is the following. Once you have this graph and can draw the graph, you could take the dataset you collected and push that through the graph, and try to visualize the flow of data through your graph by, say, having thicker lines for paths that are followed more often than other paths. The thickness of the lines could indicate how frequently paths go through that part of the questionnaire. And that may also help you detect paths that are never taken through the questionnaire and make you wonder. But it can only be done after you have done the field work or maybe you have collected some test data.

There was some discussion about text fills, and that's another important point of questionnaire instrument documentation. I have seen questionnaires where questions consist *completely* of text fills; [*laughter*] there was no question text at all, and all question text was derived from databases elsewhere, depending on certain conditions. But, how to document this? What we have done is to take the approach, more or less, of saying that every object has its own peculiarities but also every object should have some label, we call it [a] description—whether it's a question or a condition or a check. There should be a descriptive label somewhere. And you can choose everywhere in your tool to display the descriptive label instead of the question text, or the logical expression. So that gives you a way out—maybe not a very elegant way out, but an easy way out—of documenting the text fills. But [implementing] that remains a difficult problem.

What we also do not have in TADEQ is the screens. Some people say that the screens that the interviewers see should be part of the documentation. So maybe it should be possible to click on a question and get the screen of the laptop computer, a copy of that screen on display. Well, it could be included; we didn't do it. On the other hand, we thought that the text screens are there [but] the screen display for every interviewer might not be the same. [You] could have many different screen displays, and what to do with that?

Another interesting thing is that—once you have this interactive system—that allows you to play with a questionnaire in interactive ways. The next step could possibly be to use it to *design* questionnaires. Why not build an integrated development environment that allows you to just put your objects on the screen—"this is a question, connected to this condition," etc.—and just build up a questionnaire in a graphical way? I think it could be done, but it would take some time.

And then one final point is that in TADEQ we concentrate on the documentation of the questionnaire instrument; we do not concentrate on the documentation of the final data set. Tom [Piazza] mentioned that also. Now sometimes there are two different kinds of documentation. If I go to the data archive and ask for the survey data file, I get documentation—in XML—but it's not related at all to my questionnaire instrument documentation. But it's also in XML, and TADEQ is in XML, so you wonder whether it should at all be possible, in some way or another, to combine that in some way—to completely integrate it, or to make links, or whatever. And it should be done, I think, in the future, too.

So, how much time do I have left?

CORK: We have a few minutes left for questions if we want to do that. And, in particular, if anyone wants to ask questions central to his slides, ask those first because we've got to make another technology switch up here at the podium.

PARTICIPANT: A clarifying point ... What if you have [a] GUI instrument that is really a pen-based kind of thing—where it's totally interviewer-directed in terms of what modules or what I'm going to do now—but there are no predefined paths at all? Will this work, and you could actually then see what the paths are, what paths the interviewers are really taking people on?

BETHLEHEM: Well, I don't know, and I can't really give an answer to that question because we don't have that kind of thing in our organization. On the other hand, I could say that if you can get it in our XML file we will document it for you! [*laughter*] The functionality of the XML language is sufficient for your application; it can be done. But maybe it isn't. Then you should have to describe somewhere how decisions are taken, and you have to document that in some way or another. Maybe, in textual form, you could do that.

PARTICIPANT: Is there a plan to make this a production-level software?

BETHLEHEM: Yes. The aim of the project was to develop a prototype, and that is what we did. And we will put it on the Web and—in a few weeks' time—everybody can just grab it from the Web and play with it. And if you want to attach your own system to it, and you get the definition of the XML language, you can make your own interface to it. There are plans to incorporate it in the Blaise system; that is one thing. But if someone else says, "I want to incorporate it in CASES," or whatever, then they can [adapt it as well]. This was really the meaning of the project, to present something to the community and they can work from there, to see if they can do something with it. [*Inaudible reply from participant.*]

MARKOSIAN: Have you considered using a proper development platform and open-sourcing it? Because it sounds like there's a large enough community around the world that would want to contribute.

BETHLEHEM: Yes, we have had some contacts, but not to a level that we've started a cooperation. But we are open for that. Because, well, one thing about XML is that everyone says it allows you to make new standards. And, that's what everybody does—makes their *own* new standard. [*laughter*] And that's not good, I think. On the other hand, if you have different XML standards it would probably not be too difficult to get from one XML to the other definition.

MARKOSIAN: I wasn't referring specifically to the XML standard but rather just to encourage people to ... by having an open-source collaborative development you can encourage a lot of development in the end product.

BETHLEHEM: Well, we learned about this in the course of this project. And we are open to that; it's not a problem at all. But there's no concrete thing to mention at this moment.

PARTICIPANT: How well does it work? And how do you decide that it's working well?

BETHLEHEM: Well, *I* don't decide it, you decide it. [*laughter*] I think if you have very, very, very, very large questionnaires then it might turn out to be a little bit slow now and then, to generate the graph and this kind of thing. But that's not something you're doing every five minutes. So I don't think that's a real problem. Also, I think that the type of files that are generated—for example, if you have a 40,000-item questionnaire in Blaise and you generate the XML, it might be rather *long*, a big file. But it works; it works.

PARTICIPANT: In terms of accuracy, is it possible for me to write a program to see that it's parsed correctly, you know what I mean? Will it go through a Blaise program accurately, do a representation, 100 percent of the time? Or get 95 percent of it right? Or, what's your feeling?

BETHLEHEM: If you generate it from Blaise, the XML file—which you can do, if you know Blaise—then it will be an exact copy of the Blaise program. One remark to be made is that not every object in Blaise is represented in the XML; if you have very special things like coding questions or calls to an external file, that is more handled like a general container and is not specified. But if you take that for granted, it gives you an exact copy of what's going on in the Blaise instrument.

GROVES: Just to tag onto that, how many uses of this have been made on real surveys? How many users do you have worldwide?

BETHLEHEM: [*laughter*] This is a *prototype* . . . [*laughter*]

GROVES: That's *exactly* what I thought you were going to say.

BETHLEHEM: So there have been some uses, and some people have tested it in the project. But it's not in real use at the moment. It's in real use at Statistics Netherlands; they use it.

PARTICIPANT: In the data description, is there the capability for long labels to be associated with the questionnaire objects?

BETHLEHEM: Yes. I was talking about descriptions, and what you have . . . you can define many languages for an item, and you can say that one language is a long label. And it can do special things for you. So we have this capability for languages, and you can do that to record the long descriptions for export to SAS or SPSS, because they're usually in a different format.

PARTICIPANT: So you could even articulate a custom series of conditions based upon the labels? So you could [almost build] from that, potentially, a natural language description of the universe as defined by the series of conditions?

BETHLEHEM: More or less, yes.

PARTICIPANT: You could generate that, based upon the . . .

BETHLEHEM: Yes, if you specify it; you have to do it, of course.

CORK: Any last questions? Otherwise, let's thank Jelke.

COMPUTER SCIENCE APPROACHES: VISUALIZATION TOOLS AND SOFTWARE METRICS

Thomas McCabe

CORK: OK . . . and now for something a little bit different. To give a different take on documentation—a different sense of generally representing an electronic document—we asked Tom McCabe to describe general visualization tools and software metrics. Tom founded McCabe and Associates, which creates tools for software development, maintenance, and testing. He sold McCabe and Associates in 1998 and is now the CEO of McCabe Technologies. Tom?

McCABE: I didn't know where to begin, which in software is not unusual. [*laughter*] Because I have—how many people here have experience developing software? OK, most everybody. And how many people have no experience with surveys? [*laughter*]

The previous Bush administration used surveys to form a Cabinet. President, elder, Bush was in Moscow, and he met with Gorbachev. And he was having trouble with the vice president at the time, Dan Quayle. So he was impressed with Schevardnadze, and he asked Gorbachev, "How did you get this guy?" And he said, "I gave him a survey. I asked him who his father's son was, who wasn't his brother. And Schevardnadze said, 'Me.'" So Gorbachev hired him. So, elder Bush said, "Look, I can't wait to get back to D.C. so I can use this survey!" [*laughter*] So he called in Dan Quayle and said, "I have a question for you, which is kind of important to your career. Who is your father's son, who is not your brother?" And Dan Quayle said, "I think I need two days on this." So Quayle went to see Dick Cheney; Dick Cheney was Secretary of Defense back then. And he said, "Who's your father's son, who's not your brother?" And Dick Cheney says, "Well, it's me!" So Quayle went running back to Bush and he says, "I've got the answer!" And Bush says, "Well, what's the answer?" And Quayle said, "It's Dick Cheney!" And Bush said, "Well, that's wrong—it's Schevardnadze!" [*laughter*] So surveys are used a lot in D.C., here … and in the spirit of that, I'm going to talk about software.

I want to talk about visualization tools and metrics. And I'll first talk a little bit about the business—what it's like to be in the business of analyzing a lot of software, in different languages, on many computers, with metrics. Then I'll talk about algorithms, and I'll give an example of how to quantify complexity. What's struck me is that there's been a lot of discussion about complexity—I think maybe intuitive complexity—but yet [finding] the point at which things get uncomprehensible, untestable, unmaintainable … [answering] that question, *certainly*, is pertinent to software. And that happens to be the particular work that I've done.

I'll give first a definition of it and show how it works, and then I'll show a lot of examples of algorithms, in increasing complexity. There's a point at which most of us [lose] sense of what these algorithms are doing, and I realize that this is a long stretch here—so we'll give a quiz, and see where people lose the comprehension.

And then we'll talk about using similar metrics for testing and maintaining as well. It turns out that, in software, maybe 20 percent to 30 percent of the effort is in development and 70 percent is in maintaining. And I'd bet that's probably the case with surveys. But yet all the attention is on development when a lot of the effort is in maintaining. So we'll talk about how to use metrics in a maintenance activity. And then we'll talk about architecture—how to take a very large system and see a lot of software on a screen, and interact with that, and how you get a sense—beyond documentation—of the shape, the quality, the texture, the layout, the landscape of a very large application.

This comes from a lot of experience, a lot of it painful, a lot of it pretty exciting. My particular background is that of a mathematician, and I spent several years developing graph-theoretical techniques to apply to software. And then we built a company. And it was based on a mathematical idea of complexity. And it grew to be about 170 people, with offices in Columbia, Maryland; England; Germany; Paris; and Korea. We analyzed about 17 different languages, so we had parsers for Ada and C++, JOVIAL, and so on. And we worked on all the major platforms except for mainframe. So we worked on all the Unix platforms, PC operating systems, and hybrid platforms. And we probably did more than 25 billion lines of software.

What we did was, first, develop and teach the theory—actually, for about ten years. We got a lot of people to use it. And then ship products that did this. So—unlike a prototype—this stuff had to *work*. It was typically used in very, very large systems; [it has] a much better chance of working with the larger systems, it's much more pertinent. So, typically, the kind of systems we were analyzing had upward of 4 or 5—or 20—million lines of code. Very, very large operational systems, [in] commercial environments, DoD, international, aerospace. It was pretty much across-the-board, as you'd find all different languages: COBOL, Ada, JOVIAL. We even had some assembly languages for real-time systems. So, it was a real operational applied kind of business.

The kinds of things that go wrong with software are manifold. One of which is that it's too complex. And what you'll often see are algorithms that people really can't understand, and when they can't understand 'em—even though they really think they've tested 'em—they typically haven't, and you typically have errors because you can't comprehend them. It would be interesting to look at that vis-a-vis surveys.

Another thing that goes wrong [with software] is that they're not tested, and typically not tested when they're complex. So, if we take something we'll look at later with complexity 1, 2, or 3, typically, when they test it it would get tested thoroughly enough because it's so simple. As the complexity increases, the accuracy of the testing decreases. And you find a point at which it's not even close to doing a reasonable job of testing.

Software often doesn't work, and the problem with it is that it's fielded when it doesn't work. And the cost of errors, by the way—there was a comment this morning about the cost of errors being 32 times once product is already fielded. In our experience it's about 300 times. You take a software product, put it out there, and when it doesn't work—comes back in—the problem is that you don't know where the error is. So you have to localize it within the module. Then you have to change the module, re-test the integration, and then do some more regression testing. And all of that would cost, typically, 300 times more dollars and time than if you had caught the error in development.

Another problem with software is that there isn't much re-use. There [have] been all sorts of initiatives, in DoD and the government and academia, to write re-usable code. And the joke in software is that everybody likes to write re-usable code but no one likes to re-use it. [*laughter*] And one of the reasons is that nobody works at it. In the universities—[for instance,] Michigan having one of the first computer science departments—most of the curriculum is about development, design teams, metrics, coding. Very little about testing, very little about maintenance. And what happens is that everybody's developing new code when—in fact—there's often code that exists that has the functionality you need. The problem is that documentation and traceability aren't there to let you get to it. We looked at it another way—and we'll talk about reengineering here as well—and we reengineered systems and found 30 percent of the systems were redundant, in the sense that they were doing the same thing. And the problem was: people putting new things in couldn't locate where the functionality already existed. It will be interesting to see if that's pertinent to surveys.

And, of course, there's the problem of standards. There are all kinds of standards in software—one of them is about complexity, *now*; it wasn't some years ago. [It] goes toward programming style; individual corporations have their programming standards. There has not been a whole lot of success in using them. And, in fact, there's a lot worse … it used to be, when projects were bigger and well-defined, that standards would have a chance. Now, you've got kids writing XML and products being shipped based on just a very loose, collaborative environment. And standards and testing are in probably worse shape.

OK, I want to talk just a little bit about the architecture of some products. Not to talk about the products, but I want to give you a sense of the activity we engaged in, because I think a lot of it is pertinent to surveys. And one part of it is parsing a lot of languages. Now, there's a bunch of them listed here, and some have been added since.[31] One of the key things of approaching this is that there has to be a grammar underlying the language, obviously, in the computer langauge. But there also has to be a fairly consistent parsing technology, so you can extend it. And the problem is that there's no way to maintain seventeen different languages. And I think that's pertinent to surveys as well; if you try to approach the problem of measuring surveys and testing them, and

---

[31] The specific languages mentioned in the slide in question are C, C++, Java, Visual Basic, COBOL, FORTRAN, and Ada.

getting their complexity, the first thing to deal with is how to parse the language in which they're expressed.

Another [part of the architecture we developed] is the database to hold metrics, reports, and graphics for display.

Of the three major activities that we were involved in, the first was quality assurance: how do you find something early that's out of control? How do you find something early that can't be tested? How do you find something early that's not going to be able to be maintained?

The second one is testing: once we have a product, how do we test it? Now, we're going to take head-on the issue of paths—in fact, I've heard a lot of discussion about questionnaires and paths and modules and how you count a path and the number of paths and all that—and that's exactly the topic we were into. One of the problems in the testing of software is that if you had a group of a dozen people, and you look at the criteria for shipping, it was purely ad hoc—it was simply that one guy sent it through testing and a manager would say, "Fine." Well, one person might be good at it and the next guy lousy at it, but that was the criterion. So we're going to talk more about a defined mathematical criterion for completing testing.

And then a third area is reengineering. In software, what that's typically about is how you take a lot of source code—say, a million lines of source code—and how do you reengineer out of that the architecture? How do you tell whether the top modules were the ones who called each other, what functionality is within what modules, and how do you see the algorithms? [Our tools helped with these efforts in general programming environments,] so it would be interesting to see if surveys have the same kinds of characteristics.

So reengineering means: how do you get back what that architecture was? I was kind of fascinated listening to the discussion about documentation, and the pains of doing it, and the experience of it always being incomplete—because in software that's *exactly* the case. In fact, I would maintain that software is worse. What typically happens, with almost all of our clients, is that documentation consisted of what we could reengineer from the source code. There was no documentation. And, in fact, when they had it, it was misleading. Because the problem was that the documentation didn't match the system; it was done at one point in time, and then when the code was built and shipped too early, the whole organization was in such a panic to try to fix it that they had no time to update the documentation. In fact, they couldn't complete the testing before they shipped out second releases of it. So what would happen is that the documentation would be four, five, six iterations out of date ... and *dangerous*. So the only documentation was what you could reengineer.

Now let me make another point in terms of the time basis. The stuff in here—the testing—had to be done in real-time. So the way it would work is, when someone was testing an algorithm there would be immediate feedback about the paths he'd gone down, the paths he hadn't gone down, the conditions that drive it, and the percentage complete. So it had to be as fast as a compiler, and that easy to use. And I'll show later what that looks like. With reengineering, that was more on a daily basis. So it would take—for example, with maybe half a million lines of code—about a half hour to parse and represent. So, for a very big system, it might take half a day to parse all the source code, show all the architecture so you could see it. Now the other thing—so that would be just the parsing to get it.

The other thing, however, about having a big system is that there is no static view that would do. You have to have, like, a CAD/CAM picture of it. And so, for example, you would say, "give me a certain record; where does that record appear in the architecture?" So it would highlight those modules. "Where within those modules has it been tested?"— highlight those. "What modules never use it?" If I take a dataset with four, five, six, or a dozen records, show me the architecture that uses that. Show me the pieces that this user is using. So there are all kinds of questions and representations that you show very interactively, and when people go to use this—or start using it—there's no way they try to change it without that, because you can't see something that big. And I think that relates to surveys—a little bit different.

I was going to say that software has to work, because you modify it, you recompile it, you put it out there, and you run it ... but it doesn't. [*laughter*] The error rates in software are incredible. But you do compile it—and in fact do run it—and the idea is to see it that way. With the quality assurance, that would occur at the beginning and work all the way through.

OK, I'm going to show a few examples, and I chose C as a language. And that's a very small—I would call it an algorithm or a module. [*See Figure II-15.*] And it says that you begin here, and statement 0 is x replaced by 3. And then here in blue is the first condition, which says that if y is less than 4, then x is replaced by sin(y); otherwise, x is replaced by [cos(y)]; and so forth. And then the flow graph—just the notation I'm going to use—looks like this. So it says—the node 0 is here, and it comes into this node that splits. It has one entry and two exits, and that's what a decision looks like. And it goes to either 2 or 3, and then to 4.

One of the things that I hope that what I'm showing you will suggest is a sense of this being a minimal representation. The problem is that you want to show data associated with this, and a lot of things associated

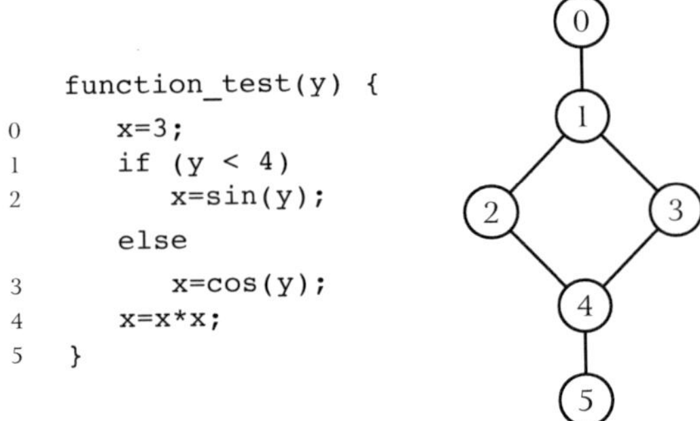

```
        function_test(y) {
   0        x=3;
   1        if (y < 4)
   2            x=sin(y);
            else
   3            x=cos(y);
   4        x=x*x;
   5    }
```

Figure II-15   Simple C algorithm with flow graph.

with this, so when you choose the way you think of this graphically you have to get a minimalist view.

Now here's another algorithm—a little more complex, and you can see the graph next to it. [*See Figure II-16.*] We're going to use some terminology here. This thing is called a "node", a node being a circle with a number in it, typically representing one statement or collection element. And then an edge is the thing that points from one node to a second. So we're going to talk about nodes and edges.

You probably can't read that, but one of the questions you could ask is—when you think about a survey, does this pertain or not? And maybe it doesn't. But it seems to me that one way in which it might is that a survey has a control flow, just like an algorithm. So, in a survey, the user would be asked the series of questions one through two and then— depending on the output—a quick exit would be out this way. And then a different outcome of the data would take you into here. And then a quick exit out of that would be around this way, and so forth. So you could think of this as being the control flow of a questionnaire.

Now another interpretation—I think, within a questionnaire—could be data flow. In other words, what this could mean is that the data at question four depends on the data at three, and one and two. Or the data I'm asking about at question six depended upon the outcomes of the data at five, three, one, and two. So it could also be thought of as a data dependency graph of the questions within the questionnaire, and it's been used that way in software as well.

OK, now here's another view. I don't know if you guys can see that. [Here's] the algorithm, and this is in C, and then you see some information here about it—the name of the module, and this indicates the correspondence between, for example, node number 8 over here, it's this clause **support()**. And this shows something about the complexity,

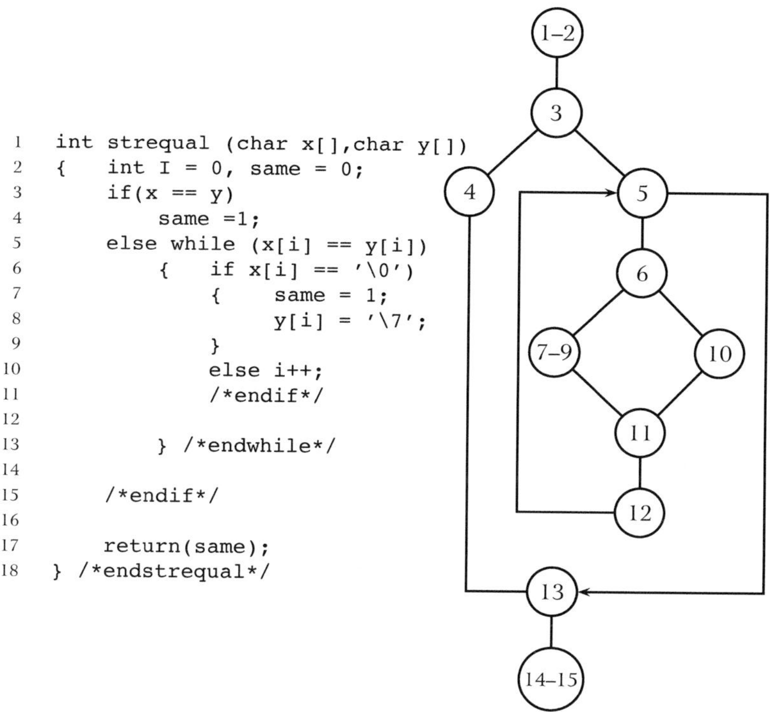

Figure II-16   More complicated C algorithm with flow graph.

and here the complexity is 2—*meaning* when you go to test that there are basically two paths.[32] The point I'm making here is that the way we're working this is to give the complexity and the graph to the user developing it, right with the source code. So there's no separation.

It's interesting with quality and testing of software ... you have to make it very, very easy. If it becomes separated or takes some time to do it, or it becomes bureaucratic, it won't be done. The point at which this particular stuff is applied is in unit integration testing, and if you don't get the testing done well *there*, the software starts to blow up as you integrate things and go into field testing and the cost is incredibly high. We've mentioned that it's about 300 times more expensive in the field; it's about 90 times more expensive in a regression or acceptance test than in a unit test. So if you're not catching the errors here, they get much more expensive later. And the idea is to make it very easy, OK?

---

[32]The graphic specifically referred to here is not reproduced in this report, and the measure of complexity is described in more detail later in this presentation.

Now let me also mention that a lot of our users would be using four, five, or six different languages—maybe Ada and C++ and JOVIAL. And all the testing, the quality, the reengineering would look exactly the same. In fact, the medium became the complexity, the graph, the numbers, rather than the syntax.

OK, I'm going to talk a little bit about three different metrics, and one is called "cyclomatic complexity" (*v*). It's the measure of the inherent complexity within an algorithm, and the way it works is that it quantifies the number of basis paths. Now, many algorithms will have an infinite number of paths—just like many questionnaires do. And the reason is you can loop. However—as a practical concern—you can't test an infinite number of things. You have to make some kind of engineering choice about what you test. And the cyclomatic number is the number of basis paths within an algorithm which, when taken in combination, would generate all the paths. There's a theorem in graph theory that says that you can think of the paths in an algorithm as a vector space— that's typically infinite. However, there's a basis set of paths that when taken in combination would generate all paths. And the cardinality—the number—of basis paths is the cyclomatic number. And we'll illustrate that with some examples. That's something I published years ago, and grew the company with.

And then what we're going to talk about as well is "essential complexity" (*ev*). And essential complexity is about—not so much the testing effort but the quality. There [are] some things in software that are inherently unstructured, and it's very similar to what you guys were talking about with surveys; I'll show some pictures of it. There [are] ways you can screw up ... in fact, I suggest that the software world was screwing this up *well* before you guys even got into it. [*laughter*] And there's a way you can quantify the essential complexity, which is the unstructuredness which leads to a maintenance trap—you can't test it nor maintain it.

And then, later, we'll talk a little bit about something called "[module] design complexity" (*iv*), and that has to do with how many modules this thing invokes below it.

All right, now the way cyclomatic complexity is going to look is that we're looking at the decision structure within a module. In a survey, I think that the analogue might be that this is a page, a group, or a collection, that's thought of as a separate group. So we're talking about the complexity within that; we're not talking about the complexity across, we'll discuss that later.

Now I'm going to show an example here. Here's an algorithm that typically would be error-prone, if this were Ada or JOVIAL or C++; it would be error prone because it's a little bit too complex. [*The schematic diagram referred to in this and the next few paragraphs is shown in Figure II-17.*]

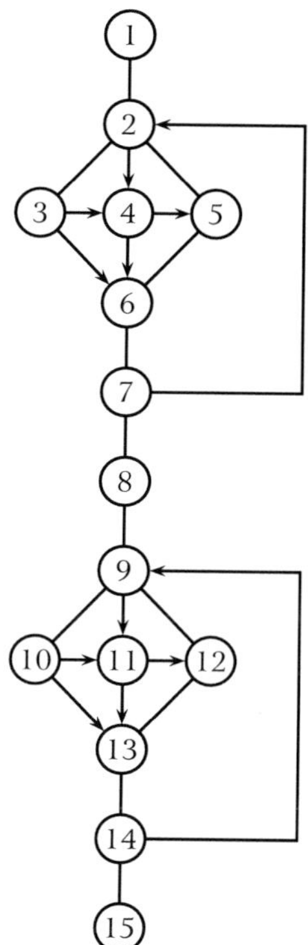

Figure II-17 Schematic diagram of error-prone algorithm.

It turns out it has 11 paths, and let me give you a sense of what they are. There's one that goes like this, there's a second that goes like this, there's a third that goes like this, there's a fourth that goes like this, a fifth that goes across here, a sixth here (around down), a seventh here, 8, 9, 10, and then 11. Anybody miss that? [*laughter*] So if you were testing that as an algorithm, you somehow want to come up with 11 paths. Now, typically, if you gave this to a software testing person you get answers all over the place. One person will tell you there are 3 tests, another will tell you there are 300. And probably every number in between.

Now let me quickly show you how this is computed, this cyclomatic complexity. This is the reason why we had the product; you don't want to be doing this stuff by hand. And one way is you count up the edges, and it turns out there [are] 24 edges. Now the edges, recall, are the transitions between the nodes. So there's one edge there, two, there's another edge there, etc. So these things in blue are the edges. And now you count the nodes—and there [are] one, two, three, four—and it turns out to be 15 nodes. And the cyclomatic number of a connected graph is the number of edges minus the number of nodes plus two. [Using] graph theory you can prove that this is the number of independent paths. And it turns out there [are] a lot of paths. Again, you get software to do this; you don't try to do this by hand. Now it would be interesting in a survey to ask the people testing how many tests they do, and that might suggest you get the same thing as in software—anything from 3 to 300.

Now another way to compute the same thing, which is a little bit easier, is to—whoops, I went a bit too fast here—is to count the regions. If you just count the regions, that gives you the complexity. So, there's a region in here—can you see that?[33] So there's one, two, three, four, five, six, seven, eight, nine, ten, eleven. There are eleven regions in the graph, and that will be equal to the number of independent paths. That's not the *total* number of paths; this thing has a hundred million paths in it, because you could iterate any number of times. But it has eleven basis paths.

You can mathematically prove that, if somebody doesn't test eleven basis paths, then that algorithm is overly complex. Have you ever seen a survey that's too complex? [*laughter*] Now what it means—and there are all sorts of software that are too complex. And what it means is that the control structure is too rich for substantiation. In other words, there aren't that many questions within it. So it turns out that if you can't test eleven basis paths—say, you can only test seven—there's an equivalent algorithm with complexity seven that does the same thing.

All right, so that's how you compute it.

Now, what happens with all this? Well, first, when complexity is high you have a high number of failures, because it isn't tested. People try to test it—they work hard—but they just don't see all the combinations. It's difficult to understand; you get unexpected results. What happens with time is entropy always increases. It always increases in software; my guess is that it probably increases in surveys. It doesn't go down; you have to work at controlling it and keeping it down. As a result, it gets unreliable.[34]

---

[33]This method assumes that the flow graph is planar—that is, that it can be drawn on a plane so that no edges cross over each other or intersect each other at any place besides a common vertex. In that case, the plane can be divided up into regions—connected areas bounded by edges that, pieced together, form the plane. In Figure II-17, each of the two diamond-shaped portions of the graph contains four regions (e.g., the triangle bounded by edges 4–5, 4–6, and 5–6 is a region). The area enclosed by the edges 6–7, 7–2, 2–5, and 5–6 (and the counterpart on the lower diamond-shaped portion) counts as a region. The area of the plane *outside* the already-mentioned enclosed regions counts as the final region, bringing the total to 11.

[34]McCabe's slides add the comment that software with complexity values $v > 10$ are "less reliable and require high levels of testing."

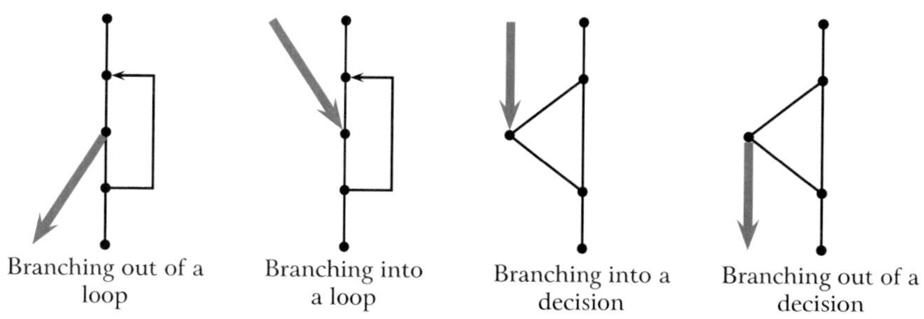

Figure II-18 Four sources of unstructured logic in software programs.

Now, I'm going to talk about essential complexity and then give a whole bunch of real examples. Now this is a little bit different; this isn't about testing, it's about quality. And it's about unstructured logic. Harlan Mills came out some time ago—with Terry Baker—with papers about structured programming, and it meant a style of logic that looked like it was pretty testable and reliable.[35] Now I published a paper some time after that about *unstructured* programming—what it meant, if you didn't use structured logic, how could you characterize the thought process behind the spaghetti code?[36] What does it look like when things are not structured? Essential complexity measures that; it measures the degree to which it's unstructured.

Now let me show a couple of examples. These are four things you can do if you want to mess up your logic. [*See Figure II-18.*] Now, you can mess up JOVIAL, or Ada, or C++—you probably can mess up a survey, OK? Now one of them is to have a loop where the normal exit is down here but in the middle of the loop you jump up. Another one is to have a loop where the normal entry is here but you jump in the middle of it. See, the reason that this software doesn't work well is that the compiler will typically initialize induction variables at the beginning and you're jumping into the body, often with a random result. Another case is where you have a decision—you have a condition, and then some result. But you jump right to the result without going through the test. And the fourth one is where you jump out of the event clause.

Now those look simple, but I proved a theorem which says that logic can't be a little bit unstructured—it's like being a little bit pregnant. If it's unstructured at one point, it'll be unstructured somewhere else.

---

[35]Baker and Mills (1973) is an example of the specific author collaboration referenced here; the ideas of structured programming were subsequently laid out in numerous papers by Harlan Mills and collaborators.

[36]See McCabe (1976).

Now let me give some examples. Actually, we're going to have a quiz, so you guys are going to have to answer these questions. I'm going to show two examples of algorithms, and here's one where the complexity is 20—it's pretty complex. There [are] 20 paths. And let's say you had to modify that node. So you have two choices—this is Choice A [*Figure II-19*], you could choose that, or this is Choice B [*Figure II-20*]. And Choice B is this algorithm with cyclomatic complexity 18, where you're modifying that node. All right, so, look at it one more time. How many people would choose A, that one? Now you've got to choose one here; you can't get out of here without choosing A or B. [*laughter*] So that's A, and that's B. So how many people would choose A? And how many would choose B? All right . . . chaos rules.

It turns out, with A, let's look at what you do. In A, if you modify it there, it would be pretty clear how you would test that, right? Because the test-retesting you would do would be down here. And you pretty much could confine yourself to thinking—if not just that statement and that path—then at least to this domain within here, or in there. So you can kind of chunk the algorithm; you can look at it in pieces and separate concerns.

Now, in the second case—say you modify it there. Where do you test it? What are the chunks? You see, it's gestalt—you either get it or you don't, and you've got to get the whole thing. And that's what goes wrong. And it turns out that this algorithm has essential complexity 17.

Now, essential complexity is bounded by 1 and the cyclomatic complexity.[37] This thing [*Figure II-20*] is 17/18ths fouled up. You see, it's almost perfect. [*laughter*] And you're going to see things much worse than this. The incredible thing is when we look to software for real, it is beyond your imagination—this is just a warm-up, as to what the stuff will look like when we get to things we're working on. Now, the one before [*see Figure II-19*] has essential complexity 1. What that means is that, psychologically or physically, I can chunk that as much as I want. I can take all this stuff and treat it separately from that stuff. I can think about *that* separately from this, and I can separate this out. There's a mathematical definition here, but if I keep going I can essentially reduce this whole algorithm to a linear sequence. So it means I can separate my

---

[37]McCabe's essential complexity is computed by analyzing the software module's flow graph and removing all of the most primitive pieces of structured logic—the lowest-level "if," "while," and "repeat" structures—embedded within the code. Once the module is reduced as far as possible through removal of the structured primitives, the cyclomatic complexity of the remainder is calculated; this is the essential complexity of the module. By nature of this derivation, then, essential complexity can not exceed the cyclomatic complexity of the whole algorithm, since the cyclomatic complexity of the remainder following removal of structured primitives can not exceed the complexity of the whole. For additional information and examples, see Watson and McCabe (1996).

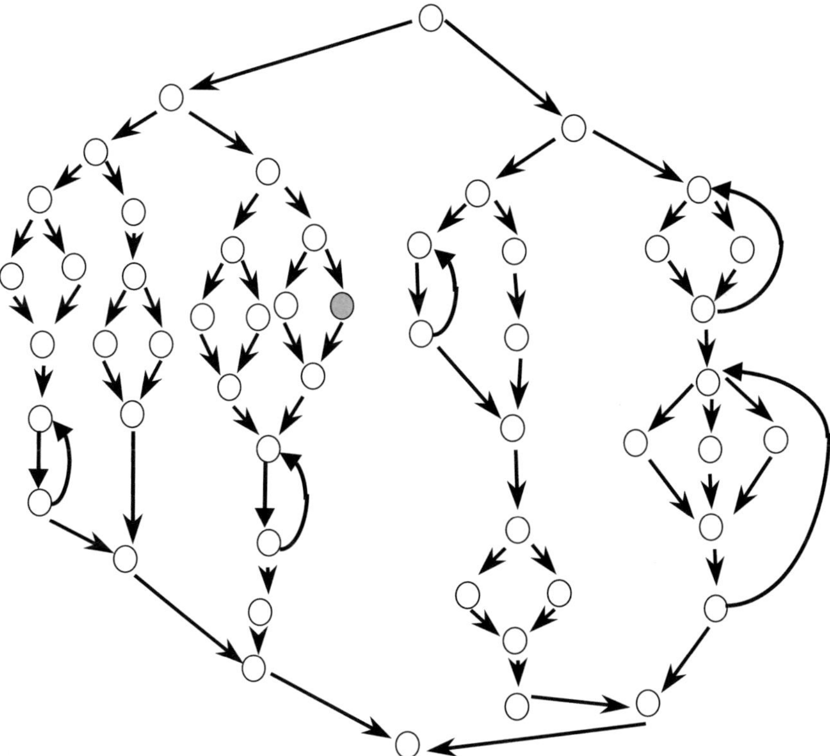

Figure II-19 Choice A in software metrics quiz.

NOTES: In the context of the quiz, the question is on the level of effort that would be incurred by making a change at the node (near the center) that is shaded in. The cyclomatic complexity for this graph (v) is 20; essential complexity (ev) is 1, and module design complexity (iv) is 2.

SOURCE: Workshop presentation by Thomas McCabe.

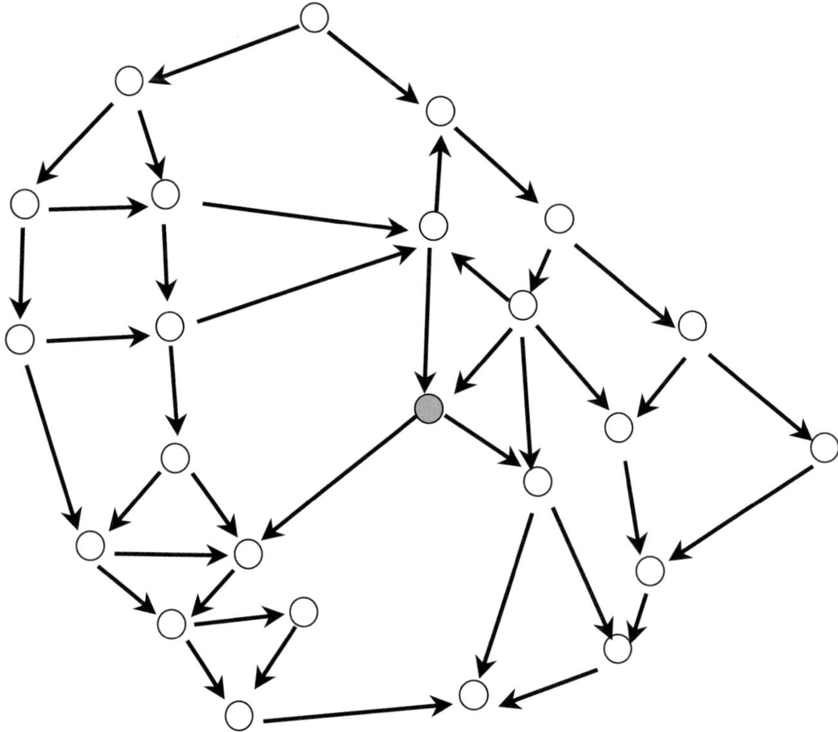

Figure II-20 Choice B in software metrics quiz.

NOTES: In the context of the quiz, the question is on the level of effort that would be incurred by making a change at the node (just right of center) that is shaded in. The cyclomatic complexity for this graph (v) is 18; essential complexity (ev) is 17, and module design complexity (iv) is 6.

SOURCE: Workshop presentation by Thomas McCabe.

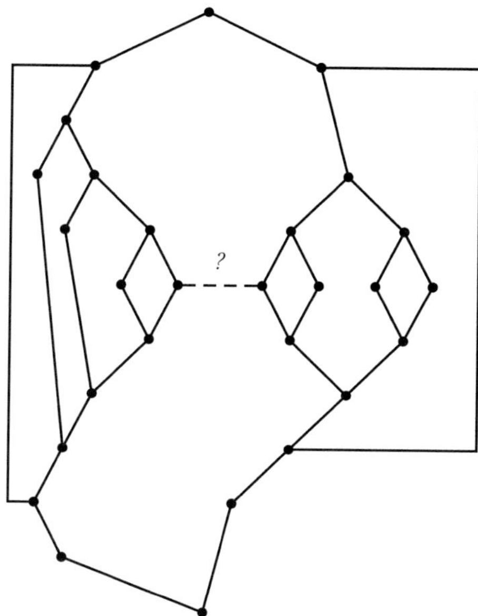

Figure II-21   Example of large change in complexity that can be introduced by a single change in a software module.

NOTES: Without the dashed line indicated by a "?" in the graph, the module has cyclomatic complexity 10 and essential complexity 1. Adding in the dashed line, the resulting flow graph has cyclomatic complexity 11 and essential complexity 10.

SOURCE: Workshop presentation by Thomas McCabe.

testing, separate my domain of change, I can factor out reusable code. I can comprehend it. Whereas in the second case I can't do any of those things. Actually, the only structured piece is down there.

Now, let me show just one more thing about this. Now, that's essential complexity. Here's an algorithm up here where the essential complexity is 1. [*See Figure II-21.*] Can you guys see that OK? And here's one change—the guy puts one "GOTO" in, from here to here—and the essential complexity went from 1 to 10. So it went from being perfect to being 10/11ths fouled up with one change. And you'll see algorithms where it's much more dramatic, cases where the cyclomatic complexity is 50 and the essential complexity is 1. One change, and the essential complexity went from 1 to 49. And that's why maintaining and changing things is so dangerous; if this were a survey, you would go from a survey where you could re-use pieces of it—you could separately test it, or modify just pieces of it—to where you couldn't do any of those things. So, small change can have big impact on quality.

So that gives you a sense of the metrics. I thought it might be fun to look at some real stuff here. I'm just going to start showing some examples of algorithms and giving a comment. And this is all real, live stuff, like we see every day.[38] And the first one, here, its complexity is 14. And it's pretty well structured—its essential complexity is 5, and the only unstructured thing is down on the bottom. But that's something that you can work with. And, by the way—we'll talk about this later— these things have function calls. So it also gives you an idea of design complexity, how much it invokes from the stuff below it. But that's something you could work with.

Now here's another example—things are getting worse here. Actually, this one has complexity 7, so it's not so bad. Now this shows the dynamic trace, which is often used in testing as well. For example, in this particular case the guy ran a test, and the test went down this particular path that's highlighted. I don't know if you can see this, but this shows that the path is highlighted down here. So it also works incrementally, as you're testing, you can graphically see what you're doing.

So this one has complexity 15, and we're getting a little bit worse. This is complexity 17. And I hope you're getting a sense that as those numbers go up that things are looking … worse. And we're not even *approaching* some of the things you see in operational environments. So this is 37. Now with the tool you can expand this, make it wider and all that. But, you know, when you have incredible complexity that doesn't help; you can just see the mess.

So this is complexity 59. Now one of the problems is that people sometimes get shocked—"*I* didn't write that!" [*laughter*] "I don't know what happened; I never wrote that! I know it's got my name on it, but … I took a week off, came back, and look what happened!" [*laughter*]

Now here's an example. I'm going to show two examples. Here's an example of an algorithm that doesn't work. It's got complexity 300. And, by the way, 30 percent of—you see, this thing, this thing, this thing, they're all the same. And if you blew this up, all of the things inside it are the same. There's just so much complexity, the guy couldn't see it. Now that's an example of an algorithm that doesn't work because it's too complex. But you can make it work by modularizing it, because the essential complexity is 1. So you can attack it piece-by-piece.

---

[38]The specific examples referred to here were presented on overhead transparencies and were not available for inclusion in this report.

Now here's an example of an algorithm that not only won't work but it will *never* work. [*laughter*] It's got complexity 262 with essential complexity 179. Now the general work in a job, the way things work in software with an algorithm like this is you go to fix it, right? And you make your first set of fixes, and those fixes introduce secondary errors. You fix those secondary errors and you've got tertiary errors. And so forth, and so forth. And it's like a bureaucracy: the more you work at it, the more work you're generating. So then you figure out: we were best off seven months ago, before we ever changed this thing—the original errors weren't that bad! [*laughter*] Did you ever work in something like that? That's the way software goes. Now, we see stuff like this all the time—like, this, you can almost see it. But we've worked with some *scary* stuff, stuff with mission-critical software that just didn't work or even come close to working.

There [are] two messages here. You can never see it without *seeing* it. You see, a lot of times the number of source lines of code is not that big. The relationship between the complexity and the number of lines was not that high a correlation. And, second, you can't test number of lines; it's the paths you test. And the other point is the metric to measure it. So that's what I'm talking about, and I wanted you to see some real things to get a sense of what it's like.

It would be really interesting to take a look at some surveys. See, I think that some of the analogues are . . . algorithms have complexity and the paths you traverse. It seems to me that surveys have complexity and they have paths you traverse. They're the same thing. And you guys were talking about keeping track of all the paths—you don't. Because a lot of the paths repeat segments of another path. So when you put the paths together you have an algorithm with an implicit complexity.

OK, and by the way, with this thing with cyclomatic complexity 262, the amount of testing might be like, say, 50. In fact, with this thing, we found the guy who did this . . . you have to be sort of diplomatic when you approach them. And he got kind of upset when he saw the picture and said, "Wow! I tested that *at least* twelve times." [*laughter*] He was *certain* of it.

PARTICIPANT: Let me just ask . . . you're defining complexity not so much as the logic problem they're addressing but their approach to programming it?

McCABE: Well, I am, in software. Like, the classical application of this is what you just said. But I think if I extrapolate to surveys I'd characterize it more the way you're thinking. In other words, I don't think that it's as you write the algorithm. The survey, like the programming language, has an existence that's fully characterized by the survey. And you get all the information from that document. And my belief is that

that document really is an algorithm—it has statements like statements in Ada, it has decisions like decisions in Ada, and it has subroutine calls which say to go around the document, go to a different area. And when you look at it together, it really is an algorithm.

[There has] been some discussion here about the paths going through it. [Someone] said that there might be thirty interviews, each one of which would be a different path. See, I don't think that would be true. There may be thirty interviews but there might be only twelve paths, because you repeat some of the paths. So I think that if you model the survey as an algorithm and then compute the cyclomatic paths, then the interview is just a tracing of one of those paths.

And, by the way, with the representation—I think that this would work with surveys. The way you would show it is you dynamically [show] both the paths you didn't test and the paths you did test. Now, when you show the paths you didn't test, you do that to do more testing—to make sure you get things right before you ship it. When you show the paths you did test, you do that to comprehend reengineering. Because, in most software systems, you spend 90 percent of the time on 4 percent of the code. Therefore, when you go through an algorithm and highlight the paths that you most go through, that would tell you guys, "This is where 80 percent of the people are going." So I think there's an exact analogy.

And then the architecture ... well, I'm assuming the survey is maybe two or three pages, whatever. [*laughter*] Well, I guess it isn't. It probably looks worse than this. There's probably a notion of architecture in surveys as well, it's another area. I guess it's like when you do the IRS stuff, the taxes, and switch to Form B or whatever you have to do. Form B is another module. And then the architecture is a linkage between the modules.

Now, likewise—I don't know if I'll have enough time to do it—but we've also published papers about design complexity, which is the interaction of all the modules. So the way you'd think about testing is, you want to know the design complexity and I'm going to force myself to do the design testing before I ship because I don't want to have the errors coming back in. If I want to do reengineering I want to see the traces across the architecture when it's being used. I think they're exactly analogous. Yes?

PARTICIPANT: I like your presentation very much, and I have a kind of orthogonal question. A lot of the new object-oriented languages—Java, C++, and so on—sort of encourage their programming to have a kind of modularity, reusability and all this. And I personally have seen a number of real-life problems faced with languages like C, COBOL, and FORTRAN really hiding the functionality rather than making it usable.

So I wonder if you have any comments about how these differences might work in testing?

McCABE: Yeah, I do. Did everybody hear the question? This is something we were talking about at lunch. What happens in software is that, about every four or five years, there's a new mantra. It was [object-oriented (OO) programming], and before OO it was encapsulation. Before that it was structured programming, and after OO it was XML. And what happens is that—when things get totally screwed up and you can't maintain them—you invent a new paradigm and declare all the other stuff unusable [*laughter*] until the current paradigm becomes unmaintainable. And then you declare another one. And what happened with a lot of our competitors—in fact, I think there's a guy in the room here—we wrote all our stuff in C, all these things, because it's very portable and we had a solid base. And some of our competitors switched to C++ about that time. And they landed up with an architecture so complex that—guess what?—they couldn't maintain it. So what happens is, it isn't so much the technology that gets us in trouble, it's the fact that—for example, in a computer software shop—you never see a line item for studying the system. Analyzing the system. There are line items for development, for testing, for coding—but not for analysis. And, so what happened with a lot of our world is that the architecture gets so complex and reusability gets so complex that there are a lot of pathologies—well, we published a lot of papers about that. For example—along the lines of the comments on Bush and Cheney—when you have the father inheriting from the son you get the wrong kind of cycle in OO.

So the new ways of thinking about the programming languages really haven't helped with that issue, OK? And, also, there's going to be a guy from Microsoft talking—a lot of some of the best quality software is done with very thorough regression testing. And a lot of the big companies in the software business do that very, very well. One thing they do well is—that when they make a change—they can pinpoint the subset of regression tests pertinent to that change. Now there's a thing which I may or may not get to about data complexity. And data complexity says, "if I change these variables, what's the locality of them in my architecture? And what's the data complexity of the change? And give me the regression tests that pertain to that data ... not the other ones."[39]

---

[39] McCabe's (unused) slides for the data complexity metrics define two such measures. The first, global data complexity (*gdv*), is said to be "a measure of the usage of global (external) data within a module" and "is associated with the degree of module encapsulation." Hence, if two modules had equal cyclomatic complexity numbers but different values of *gdv*, the one with lower *gdv* would be preferable due to its stronger encapsulation. Specified data complexity (*sdv*) measures "the usage of local data within a module," so that high values of *sdv* are a positive design feature.

So therefore, what happens—you take a regression test that maybe has 14 million tests and you can pinpoint the—maybe—dozen that hit the things you're changing. Now we used that in the year 2000, because the year 2000—the year, the data field—was driving the whole thing. But you find often that the organizations producing quality software are getting it not so much as the result of a new paradigm. It's by engineering principles, stuff like regression testing and the stuff we're talking about.

Let me just do a little bit more here—how am I doing with time?

CORK: The good news is that we're going to try to copy Tom's slides—and everyone else's—and have them to distribute tomorrow. The bad news is that I need to ask you to wrap up in about five minutes.

McCABE: Let me first take questions. Yes?

MARKOSIAN: This question isn't directly relevant to the application, but more of a historical question. [*inaudible*] And then things got better over time, and now there's been remarkable development that seems to be sending things back to the beginning, which was to proliferate current systems. What you're doing there is introducing two things: the enormous complexity, the syntactical kind of complexity that's represented in your chart. And also that complexity is hidden because it's not available for the programmer to look at, things like leaving to the operating system. So, do you have any approaches to that?

McCABE: Yeah, there are a lot of them, and some are not pertinent to the subject here. There are a lot of just engineering principles, about testing and project management, that pertain to that. And there are collaboration tools that help with that. So there's no silver bullet. But within that I would suggest that, within the newer systems, these issues are even more important because one of the facts of life these days is that you get different contractors and people separated geographically in different countries all working on the same thing. And, [for instance], what Microsoft does, when you recompile every day, the idea is to visualize the thing you have. Easily and frequently, and share that. And it's probably more important *now* because of what you say than it was back then. And things are growing in complexity, not the other way around. And the environments are more complex, in fact, and hearing the standards is even more important.

So I think—let me just summarize by saying that you're saying that we want to know things about data complexity, and that's a way to view a subset of data, and then what complexity it induces, and how to test that and how to change that. And a lot of this is about visualization. It was striking to me as we discussed it this morning that the issue, it seems to me about surveys, vis-a-vis documentation and testing, requirements being together up front—are entirely analogous to what has been happening in software. And in software the joke is that you never get to requirements; the real joke is that the requirements end up being what the source code does. [*laughter*] And there's a certain pathological truth to that, but that's the way you end up. And then you re-engineer your way back out. And the other thing that happens all the time in software is that the first thing that goes is documentation. And right with it goes testing. And documentation winds up being what you can get out after the fact, and the testing—unfortunately—has been very ad hoc, at incredible expense. Now, one of the things you're going to see in the handout is that we developed some technology that said when you think about your architecture early, there's a way to figure out your integration test before you build the system. In other words, you can take the way you are thinking about the system and build the high-level test before. And that's the only time you get the leverage, because once you start building it there's such a panic to get it out that you can't [get] in the middle of that. So the real psychology is: how do I develop my test before the fact? And there it's clean intellectually as well because you know ahead of time what it should be doing. It's not that you follow what it's doing as the requirement.

So I want to thank you. I've kind of enjoyed this because this probably sets for me the record of not being a bird of a feather, if you know what I mean. [*laughter*] Because I have no idea in the world, and I promised myself when I got here not to study it because I wanted to come out of the box clean [*laughter*] with no preconceived issues. But it has stuck me that the issues I think are very, very similar. So thank you.

CORK: We will take a fifteen minute break. In keeping with the NRC policy of constant feeding [*laughter*], there are cookies and such in the back. We're about a half-hour off schedule so hopefully we'll make up a little bit of time. But we're doing okay.

MODEL-BASED TESTING IN SURVEY AUTOMATION

Harry Robinson

CORK: If everyone could settle down, we have a last little stretch here, but we are in the home stretch—the last segment for the day. The last segment that we're going to do today is focused on the problem of testing of instruments. In particular, it's going to be centered around the idea of model-based testing, a computer science approach that seemed sort of immediately applicable to some of these problems. And, to give a general idea of what model-based testing is, we have one of the best practitioners of it here, Harry Robinson from Microsoft.

ROBINSON: Thank you, Dan. Can everyone hear? Okay.

So, yes, like my confreres from the software world, I know nothing about surveys. So let's just get that out right away. But what I do know about is testing, and from what I see you folks are leaving the world of surveys to some degree and moving to software, and so expertise on software testing may inform your efforts to test your questionnaires.

So I work for Microsoft; I'm currently with the Six Sigma Productivity Team. I'm in charge of test productivity initiatives and, before Microsoft, I was with HP for three years and before that with Bell Labs for ten.

OK—the theme. I test software, you test surveys. But your surveys are becoming software. And there [are] actually models that underlie them both. And what's been fascinating to me is that even though I have nothing to do with surveys during my normal life the number of things that people have said about how you do surveys keep resonating with me and how we do software testing.

Once you left the world of paper, you entered the software world. And in much the same way that we are very feature-driven in the industry, you folks sound like you're feature-driven because you *can* do it. And so what you're ending up with is the same kind of tension we have between features that you'd really like to get in and your ability to validate that you haven't introduced bugs into your system along the way. Lots of bugs, evidently, from survey instruments; it sounds like a lot to me.

Just in terms of the cost of bugs, just for a second, in keeping with the trend, $300—I mean, 300 times—is a low estimate. [*laughter*] Because if you find a bug in the field—which is part of the reason why Microsoft really doesn't need the reputation of releasing things to have people test, because once you've released it, by the time you get it back you have to regression test it through all those systems. And then you have to redeploy. Oh, it's terrible. So let's use 300 as a low estimate.

And another thing is that I'm fascinated to hear the work on complexity. I know of a project before my time at Microsoft where—if you're familiar with software bug trends—usually the bug trend towards the early part of a project starts up *here* and then comes down *here*, and you kind of say, here's where we're going to release. There was a project where it was essentially a big bowl of mud and every time you fixed something you broke more things. So what they actually had—so, *this* is basically a zero-defect type approach … what they actually had was an infinite-defect approach. [*laughter*] Every time they fixed it they broke more stuff, and they did what was really the sensible thing—they shipped. [*laughter*]

Software testing problems—[first,] time is limited. You have some time—in between when there is something to test and when you actually have to give it to somebody—[in which] you need to be able to find as many bugs as you can. And what you need to do is address your efforts all along those areas of time to be doing the best things at the best times. Applications are complex, and getting more so. Used to be that you had your spreadsheet, you had your database. Now your spreadsheet talks to their database and in fact is callable from within their database. Requirements are fluid. "Fluid" is essentially a euphemistic term here, although I love that clip art here because it looks like they're swimming upstream.

So, some ways that we've tended to handle software testing in the industry ... Manual testing is kind of the first thing everybody thinks about, OK? You sit down and you actually try it out. Then you sit down and try it out a slightly different way. And you do it again a different way, and you find some bugs—this is good. You give those bugs to [Development], Dev fixes those bugs, they give 'em back to you. You try it out this way, you try it out that way. After about ten iterations you are glassy-eyed and just banging at keyboards and couldn't see a bug if it bit you. It's very labor-intensive and it's not very useful; actually, it has a very short half-life. You find a lot of superficial bugs but that's about it.

Scripted automation—now this is what we were talking about. This is where you take a path through your questionnaire and then you save that, so that you can re-play it later on. [I'm referring to it as] scripted automation; it's also called something like capture-replay. It sounds like it's good, but actually it is a bit of fool's gold to you, because what it does is—you are doing scripted automation, picking your way through the questionnaire. You're actually looking for bugs, but by the time you get that automation path working, that automation path has pretty much found all the bugs you're going to find. So you are not getting the benefit from that. What you're actually finding are regression bugs, but you're not going to find regression bugs in that path because, if you change that path, then your script doesn't work any more. So what you're actually doing is finding bugs somewhere else in your process. So scripted automation has its place, but not to the degree that it gets relied on.

So let's look at models. And my definition of a model is very general. A model is a description of how the system behaves, on some level. So think about what people were saying earlier in this; you know, "we lost the point where we could get our head around it." Well, models are all about keeping your head around it. I have a model of the way my car works; I know nothing from under the hood. But I know that you turn the key and it goes, "*Vrooom*," and that's good. You turn the key and it goes, "Vrunk," and that's bad. And sometimes I have to go find someone who has a better model of what's inside my car [*laughter*]. But the thing is that I could actually test my car, because as long as somebody knows the difference between *"Vrooom"* and "Vrunk," I'm there. To the same extent, if you can explain to your computer something about how your system is supposed to behave, you can use the computer to generate the tests. And for companies in the industry this makes lots of sense because computers will work even longer than new hires [*laughter*]; you don't have to pay them, and you don't have to give them stock options. So to the extent that you can push your work over on to computers, it's good.

I *love* the fact that everybody seems to know what a "graph" is. [Usually,] I have to say that it has nothing to do with x [and] y axes. [The "graphs" we talk about have [a] start node, end node; an arc is how you get between—also called an edge. So, typical kind of graph, OK? Nodes, edges. And these are unidirectional, too, by the way, so there are little arrows saying that you go from here to there.

You can model survey routing as a graph. Now, there are other ways you can use models in testing, in testing the questionnaire. You could use models, for instance, in testing data validation on the various questions. But what I'll deal with here, really, is using models to test the routing.

So, here's that same one but now we've given some names and meanings to what's going on. So here's my questionnaire. [*See Figure II-22.*] I start my questionnaire; first thing I ask is age. If the age is less than 17, I'm done. If the age is more than 16 I go to marital status. At marital status I ask, you know, single, married, divorced, separated. Single, I'm done. Otherwise, I go here. Once I finish that I ask spouse age. And so this is very simple but I think it's not . . . I read Jelke Bethlehem's thing, and reproduced it, and actually got part of it wrong. [*laughter*] Why I'd want to know the spouse's age when I'm divorced, I don't know.

So here's how you would tend to think about testing this, a walk through the graph. {Start}, {less than 17}: that's one test case. {More than 16} and {single}, another test case. {More than 16} and {married} . . . well, now I'd like to know your spouse's age. {More than 16}, {divorced}, {spouse age}. {More then 16}, {separated}, {spouse age}. So I've tested every possibility there. So that's what my test cases look like, that I'm now going to stick somewhere.

Now let's do something that I know a little bit about—the clock.[40] You know, it no longer even ships, which is sad. But the clock—everybody knows basically what the clock application does. It's pretty simple. It's actually kind of complex in some ways; it's got some intriguing bugs. And it's actually kind of hard to test because, for instance, it's hard to tell what time it is for your automation.

---

[40]This refers to the NT Clock, a Microsoft Windows accessory program that displays the current time in digital or analog format.

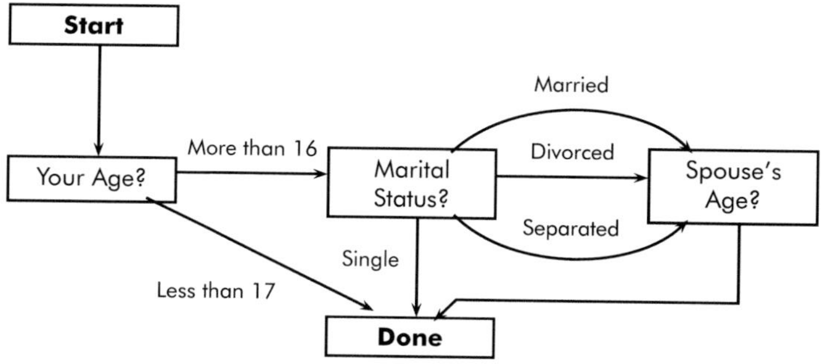

Figure II-22 Simple survey example.

SOURCE: Workshop presentation by Harry Robinson.

The clock is a graph. [*See Figure II-23.*] I could talk about an application about having states as well, just as you can talk about being on a particular question. I can talk about, well, now, I'm sitting here [at {clock not working}]. I {start} it and now I'm running it and I've got the {analog}, which has the hands. If I switch to {digital}, now I'm over here. If I turn it off, I'm actually not running, but I'm now not running in digital mode. Because if I start it again, I'm here. If I go {analog} again, I'm back. Interestingly, if I go from {digital} to {digital}, I'm back to {digital}. I don't know why that's enabled as a possibility, but it's there. But what you might think of this . . . well, I'll get to it.

If I were going to test the clock, here [are] my test cases. First test, I {start} it and I {stop} it. I check that into my test case manager. I {start} it, and I make sure that it goes back and forth between display modes [(from {analog} to {digital} and vice versa)]. I check that that bit about being stopped and coming back in digital works. And then I check one that actually runs and picks up those little self-loops. Now that is my test case group.

The problem with it is that it is a bunch of separate paths that I am saving, OK? They're hard-coded, and I'm going to have lots of them. It's not unusual for people to have 10,000 test cases like this. And it only does what I specifically made it do. I took it through it by hand and now it's going to repeat that. But this is what, in testing, is called the "minefield fallacy." The best way to avoid stepping on a mine in a minefield is to follow somebody's footsteps who made it all the way across. The best way *not* to find bugs in a software program is to run things that you've already run. OK, so now you're maintaining this, and everybody

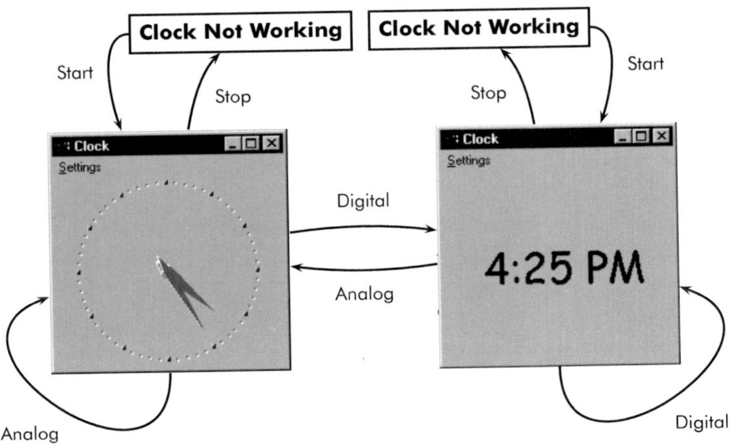

Figure II-23 Operational states of Windows clock application, viewed as a flow graph and a model.

SOURCE: Workshop presentation by Harry Robinson.

is usually proud of having a big test case database. The problem is: now, they're wearing out and, worst of all, somebody changes this functionality. Now you have to go back and re-record maybe hundreds, maybe thousands, of tests. But wait a minute—that is exactly what you are doing here with the questionnaire. Same sort of a problem. I don't know how many automated test cases you have, but if you have automation you probably have a lot.

OK, Twinkies.[41] I have no idea what the shelf-life of a Twinkie is. But I can tell you that the half-life of an automated test is probably a few minutes. By the time you've checked it in to your test case manager, it is not of much use to you anymore. And it's even worse because once you've started migrating over to different systems, different test environments, then you're porting code that isn't much use to you anymore.

So, go back to the model; go back to the description of what the system does. And let that model generate tests for you. So, this is why I'm fascinated that this workshop is about documentation *and* testing, because what you'll see out of this—hopefully—is that documentation and testing both have a lot to do with each other. One thing models are about is about being executable specs; they are documentation that you can actually use to test the system.

---

[41]The slide shows an image of a Hostess Twinkie snack cake and reads "Robinson's Twinkie Law: The shelf life of an automation script is significantly less than the shelf life of a Twinkie."

OK, so here we are with this—don't worry about how you would actually represent that. There are tools that allow you, in a simple way, to represent state graphs. Big state graphs, sometimes.

You can do a random walk. Now, the nice thing about a random walk is that it will give you unusual things that you might not have thought of to try. OK, so, maybe there's a bug [you are unaware of; for instance,] if you hit {analog} three times in a row. [If] there is, the monkey [might pick it up.] This, in Microsoft terms, is partly called a "monkey" Because it's like [the adage that] if you put enough monkeys to typing that eventually they could produce Shakespeare. This is better than a monkey, though, because what it does—monkeys actually just hit the keyboard and wait for something to crash. What this does is, as it's running, it can say that I know I'm in {clock not running} mode. I do a {start}; now I know that I should be in {analog} mode and I should be running. And you can write your automation to verify that. And then what can you do now? I could do {analog}, I could do {digital}, or I could {stop}. You do it; once you've done it, you can verify the answer again. So it ends up being a very powerful test for you. Stop me if there are questions, by the way.

Testing every action ... in graph theory this is called a "postman walk." Because it's like what a postman does delivering the mail. If you think about these as being streets, the postman has to visit every street. It may be that a postman has to walk down a street he's already walked down, and that you try to keep to a minimum. But what you want to be able to do is to exercise, here, all of the actions in that state graph. So, you're executing all the actions. And at the end of all of those executions, you're verifying that it came out right.

You can say, I don't really care about little loops; what I really care about is where I think the bugs are, and I think bugs come about when I change state. These aren't changing my state, so I'll save them for later. This is a big thing in models, when you're using them to generate your tests. You can use them to generate millions, *billions* of tests. What you want to do, first, is generate those that you think are going to find your bugs.

Here's another one. Find me, starting at {clock not running}, [every] path through this graph that has length 4 or less. And execute it. OK? Now, this, we had somebody who was doing an [Application Programming Interface (API)], a function, and actually a cluster of functions.[42]

---

[42] An application programming interface (API) is a set of conventions and protocols by which a software application program can interact with the computer's operating system and other services. It provides the building blocks by which programmers can construct software that is consistent with the operating system but in a manner which can facilitate portability of code.

And her management felt—actually, her word was "cringey" about the fact that they were just doing to do random walks through this API. So what we did was to sit down and figure out an easy way to implement this algorithm on this API. And what were able to do was to get every path through the states of that API of—I think it was 12, 12 steps or less. That's a whole lot of coverage. It took a lot of machines, but they had a lot of machines; it took some time, but they had a week. [*laughter*] I'll demo it at the end, if the demo deities are smiling on me.

OK, back to surveys—here's that survey model again. Maybe what I'd like to do is random walk through the survey. I [add] in a pseudo-transition here that's a dotted that line that says, "when I'm done, go back to the start." So now I could have somebody, I could set this off to run and it would say, "OK, now I'm going to start the survey and now I'm on question {your age}." OK? From question {your age} I can verify whatever I can about your state. Then, because I'm the one driving the test, I can say, "I know, from my description of the system that if I give an age of more than 16 it should go one way; if it goes less than 17 it goes the other way." I put either one of those in, and then I can verify that I did get to the state that I want. The problem with a random walk is that, again, it's going to do things like, for instance, here, it's going to go through the same path twice in a row. If you don't have an incredibly large space, though, this will eventually hit everything. The problem is that it may take a *while* to hit everything. Yeah?

BANKS: [Very] often, things will go wrong [in areas where] they ask for your age and somebody will say "twelve to sixteen." Or they will give it in months, or they will say "16 and a half," and it's not clear to me that this structure you've laid out captures that.

ROBINSON: So, what do you do, how do you handle error cases, essentially? Right? Where somebody gives you an alphabetic input on {your age}? In this module, you could actually—in the same way that {analog} curves back onto itself—you can say, "if I have an invalid input, then I will process the invalid input and come right back to the question." So you would end up having little loops along there.

One thing there that I thought you were going to ask is what if they were something like, "OK, I'm on {marital status}—oh, wait, I forgot, I'm less than 17." Well, this whole thing of backing up—that would be *awful* to do, trying to get coverage in scripted automation. But models could handle that fairly well.

What's interesting, then, is that models make you think out your questionnaire. So, for instance, one thing you can do in models is to check for internal consistency. For instance, say that I am less than 17; then I will expect not to see a marital status just given this. You know, it will be a "doesn't apply." But that changes if I allow going back. Because now somebody can come along and say, "start, more than 16, divorced, back, back, 15 years old." So what it does is say to you that you could put a "back" in there, [and] this is what's going to happen. And there are ways that you can now help out—from the testing side—even before there's code or even before there's a questionnaire. Because you know these are the ramifications of what's going to happen.

OK, now here's a Chinese postman walk—check every answer. Now this is particularly bushy here because we had to pick up all of those. Now, with a bigger questionnaire you could actually get in a lot of overlap. So, for instance, I'd be interested in the correspondence between a postman walk and the basis paths; it seems like a there's a strict correlation going on. What if you just delivered a questionnaire to me and we need a really *fast* thing that just says: are the questions right? Did the questions come up correctly? I don't really care about the answers yet. Here's something that will take me through all the questions, so this is actually a travelling salesman's walk. So I might want to run this before I run the one that checks every question plus every answer.

How many paths are there that are less than some number of questions? So, for instance, my kids are about this big now. So we were throwing some stuff out, and I came across "Chutes and Ladders." And if you remember "Chutes and Ladders," it's where you count up and go up ladders and go down slides ... so I modelled it. Do you know how many, what the fewest number of moves is for getting through "Chutes and Ladders"? *I* may be the only one ... [*laughter*] It's seven. Do you know how many ways you can get through "Chutes and Ladders" in seven moves? [*inaudible reply*] Wrong; 438. OK, so if you had to test "Chutes and Ladders"—which is kind of like a questionnaire—you know, you go here and then you're there—yeah, you could actually say to it, here's what it should look like. You have thrown a six, and now you're here and you should end up down there.

Here's for the Markovians in the family. If you know the likelihood of answers—[for] instance, you know where people are most likely to go in and give you various answers ... [Suppose you take a guess or use previous data, and] you've assigned a probability to all of those links. [*Referring back to Figure II-22, the hypothetical probabilities on the presentation slide are: .7 for age less than 17 and .3 for age greater than 16; and .3 for married, .15 for divorced, .05 for separated, and .5 for single.*] That means that every path from start to done now has a probability associated with it. You can figure out the path probabilities and sort them, and you can sort them so you hit the most common paths first. So, for instance, 70 percent—you pick up 70 percent, 30, here's another 15 percent, and 30,

30, another nine percent.[43]

[An advantage of this kind of testing is that it can reduce the number of steps to reproduce a bug. For a particular path,] if it does repro the bug, then I have a new shortest path for repro-ing. For instance, here's the bug. Let's say I chose "clock not running" and "digital but running"—that's the shortest path, since the path would now go "start, digital, digital, stop," and it would hit the bug, so that's my new shortest path. Do it again, and that's my new shortest path. Yeah?

PARTICIPANT: Tell me—when you're using these models, who's doing the actual testing?

ROBINSON: Who's doing the actual testing? You'd have to rephrase the question to, "Who is actually executing the tests?"?

PARTICIPANT: Let me give an example. Suppose that I am the questionnaire designer and I didn't want to ask the spouse's age if the person was divorced, but the programmer programmed it to do that and they thought it was supposed to do that. So, if I was the person who was actually running the questionnaire, looking at the output I would not want to see an age and a spouse's age if marital status shows divorced. I would know something was wrong, but the programmer wouldn't.

ROBINSON: Right, so what—there's a couple of things you could do there, and that's why I asked what you meant by "testing." Because there's testing that a tester does to understand how to put together the model but the actual execution is done by computers without people being hands-on. People can be hands-on for other stuff. So, for instance, if you know that you don't want a spouse to have an age—excuse me, divorced spouse—if you don't want a divorced spouse to have an age, you could actually put into your model checker something that says, "if I have divorce and an age, then something has gone wrong; I've gone down a wrong path to an invalid statement. I shouldn't have gone that way." It gets more complicated when you allow people to back up because now they could have gone down and said, "yes, I'm married, here's my spouse's age, back up, back up, divorced." And somehow worked it in there. So you'd have to be tricky about that. But part of it is the tie-in with documentation, OK? If the spec says that you shouldn't have a divorced spouse with an age, then that actually becomes part of the model. And if I say "divorced" and it comes and gives me a spouse's age, the model is going to expect not to get a spouse's age but to be done. And so it will say, "I'm on the wrong question."

---

[43]To clarify: the single path {start}{less than 17} has associated probability .7; the path {start}{more than 16}{single} has associated probability .15; and the path {start}{more than 16}{married}{spouse age} has probability .09. Hence, the coverage using these three paths only is .84.

PARTICIPANT: Yeah, yeah—I understand what you're saying. But what I'm saying is that it's not the same person. Is it the programmer or the person who understands the questionnaire? And, just like the person who's programming the clock, is it the person who understands the clock or the person who's been given the spec?

ROBINSON: The person who's been given the spec—this is black box test, so it's independent of how it could have been implemented. So, for instance, I don't know whether you've implemented your clock in C or C++ or COBOL—all I know is the external behavior that I expect from it. And then I can test to that.

DOYLE: In that sense, are you suggesting that we basically write the instrument twice? Once, say, in spec mode, and once to test the logic? I mean, is that what happens here?

ROBINSON: So, the question is: are we actually writing the instrument twice? And, in a sense, the answer is: yeah. In the same way, though, that I am creating a model of the clock that doesn't actually have any moving parts. They have written the clock in C or whatever; I have written the clock as a small, finite state machine—a small graph that's moving around. So what I've done is created a model that's much simpler than the actual clock. OK, say, for instance, for doing the data on the questionnaire, you might have a whole bunch of database operations that you have to go through and all this. You need not model it that way, though; you can model it as a finite state machine. It would be a *big* one, but you are doing a different sort of implementation than they are . . .

DOYLE: And what sort of language are you implementing this alternative version in?

ROBINSON: Well, these can be implemented in anything. Some of my models are done in Visual Test; there are some tools on the market that allow you to specify. I would guess—and I don't know—but I would guess that the questionnaire definition language could be used, because it is a model of what's going on. Because by reading that XML . . .

DOYLE: So, for instance, the current document . . .

ROBINSON: No—for instance, about the . . .

DOYLE: TADEQ?

ROBINSON: Yes, the TADEQ thing . . .

DOYLE: But that's what's coming out of the code that's been written . . .

ROBINSON: OK, but what you are doing as you are going through . . . you are saying that if you are generating data that is, oh, not for this one, but more than 18 years old, then I should know what it is that is going to happen. So you are doing what is equivalent to, basically, two-version programming. But the trade-off in performance that you do and what actually has to happen is much deeper than what the model is going to do.

DOYLE: Is the level of effort required to write the second version—equal, half, one-third of the effort needed to write the original?

ROBINSON: OK, so what's the comparison of the effort levels? I have a quote on this; it's about 10 percent, because what you're doing is modelling the fairly straightforward parts, and you're not doing stuff like database accesses, and you're not doing communications.

PIAZZA: Could you do some of this evaluation after the fact? [Particularly] thinking of these big questionnaires ... for each of these possible questions you could choose one [answer] at random, and how bad is that answer, and follow where that takes you. Answer that at random, and so forth ...

DOYLE: That's the random walk model?

ROBINSON: Yes, that's the random walk. But the difference is that when you do these sorts of random walks, it may look random to the questionnaire but you know what you're choosing.

PIAZZA: Right, yes, but that's the hard part! I'm kind of raising, or promoting, at least the possibility of generating datasets—the results of random responses to the questions, and you could get 50 or 100 thousand of these. Produce the dataset and then start looking, I suppose, for the sorts of anomalies you're talking about—somebody who's not married but has a spouse's education listed. I think that what you're talking about is doable, but I'm asking whether this approach might be more efficient than programming a parallel instrument, where you're evaluating at every corner whether it's doing the right thing.

ROBINSON: So you wouldn't need, necessarily, to evaluate at every corner; it works out nicely, for applications, because applications don't necessarily have an end. I mean, you could keep coming out with iterations on the clock all the time. It may be that you have some notion of what you want the answer to be; then you could check it at the end.

OK, this is a—I can do this at the end, if we have time. I don't actually own a watch, so I don't know how we're doing on time ...

CORK: About ten minutes ...

ROBINSON: Ten minutes, OK. This is a bug in the clock. It's actually Y2K-compatible—it didn't work before year 2000, and it still doesn't. [*laughter*] What actually happens is that you go through a particular sequence and the year goes away. And what you can actually do is, in the demo, there's an 84-step sequence that does this. You can just start this automatically going and it cuts it down to, I can't remember, six or seven steps. And it's all purely automatic. This, to us, is a big time-saver, because we have plenty of machines that can bang away at shortening repro sequences. What we don't have is a lot of people.

OK, so for instance, the random walk you could do much the same. And it would just be easier to find out: where exactly did we go wrong? Regression testing—part of the problem with regression testing is that you don't—you know when you give a bug to a developer with a repro sequence, when you get it back you know that the *one* thing that the developer checked is the repro sequence. OK, so it's almost useless to test that; what you'd really like to do is to test things kind of in that same neighborhood.

So, for instance, what you could do is to take your graph and assign weights to the links. This is sort of a poor man's Markov chain. Make the weights along this chain much less, and say: find me paths through the system that have as low a weight as possible. What you're going to end up with are paths that kind of stick here but that do go up and down, and around. So you begin to evolve—for instance, if you have 5 on the thin ones and 1 on the thick blue ones, that would weight 4. That would weight 18, another 18, 20, 20, and so what you end up doing is kind of generating this cocoon of regression tests around the area where you found the bug. This is actually more useful to you because you care about things in the periphery and not *exactly* where the bug was found.

So, for instance, if you found a survey bug there, you would probably test that *one*. But what you want to do is look around that one and say, give me the other things that go through somewhat similar paths; that's what I'd really like.

OK, so here's that quote: 10 percent of the program code is specific to the application. And that's usually the easiest 10 percent. So you're dealing with a much simpler problem. And, for instance, for here, what we run into is: you have to get a response back to a user in half a second, two seconds, something. And so as you're moving along here to get faster and faster with speed, you're choosing more and more complex algorithms to use because they're faster. For the testing, we don't care that much about the speed; we don't have to go through something *that* quickly because we don't care that much and because we have control over the data, we don't need all that complexity. So where the testing comes in handy for us is, we are doing these two implementations but there's a differential here. We're trading something we don't care about—which is the speed at which something runs—for something that we do. We have a better chance of getting this model right than we do of actually getting the system under test right. But they should get the same answer, and we can run them against each other.

Early bugs—you find a lot of bugs in just the spec, just sitting down and saying, what happens here? What happens there? We go completely opposite—somebody earlier said that you shouldn't automate, things aren't stable enough in version 1, or something. We actually automate as soon as there's a feature, something into place, and then we incrementally grow it. So there's something like the "turn clock on-and-off" sequence in there because you can continue to grow later functionality from there.

Easy to maintain, lots of tests—tests of what you expect and what you don't. You can put in enough oracles, test oracles, things that will tell you when things are going right or going wrong so that you can find bugs that you wouldn't necessarily be testing for directly.

PARTICIPANT: [*Inaudible*]

ROBINSON: Oh, I'm sorry ... "pesticide paradox" is the thing about a test wearing out, when you have a field full of bugs you spray it with some pesticide, you kill 98 percent of the bugs. But the remaining 2 percent are resistant, and you're not going to get them no matter how many times you apply that pesticide.

This is not test automation that's common in the industry yet. It's a difficult sell—let's say it that way—because it's much less obvious that you can get benefit from it. It benefits largely from things that you are going to do over some amount of time. So it doesn't have the flash that a capture-replay has. So what happens is that you end up with testers who have learned the other way and are uncomfortable with this or for whom this is outside their skill set. We have lots of testers, for instance, who are liberal arts majors and don't program at all. This can make them uncomfortable. You need testers who can design, who can understand what the thing is supposed to do.

It can be a significant up-front investment, finding out in advance what's going to happen. The notion we go with is that it is an investment, though, and what we tend to do is to start small and let the models pay for themselves along the way, because they keep the product very stable. It will not catch all the bugs; I was helping the people on Xbox do some modelling for their stuff, and somebody said, "you know, we have this problem where the elves sink into the floor." [*laughter*] I don't know a model that can keep your elves from sinking into the floor; what it can do for you is to verify that you're navigating through the game, or whatever, correctly.

Metrics, they had metrics—bug counts. Bug counts were *always* a bad metric. You know, "we found 10,000 bugs." Is that good, is that bad? Because nobody cares how many bugs you found; they want to know how many bugs you didn't and left in. Number of test cases—if you can generate test cases, then the number of test cases gets a little bit iffy. You can say how much water there is in your bucket, but how much is there in your faucet? Better metrics—coverage. Have you covered the spec, the requirements? In the code? Have you covered the model? So things where you can say, "this is where we are; we've covered this much. Here's the amount that we don't currently know."

PARTICIPANT: So what type of testing do you use, to supplement the model-based testing, since that doesn't catch all the bugs?

ROBINSON: We use some scripts, because scripts are very good for doing build verification tests, tests where you've just got to know if this particular part runs. Sometimes, though, we generate those with models, because there's just something that you want to get going, get running. What this does, though, is it frees up people to do the kind of testing that people are better at—people are better for judging that something looks right, that it's got a good feel to it, that it's easy to use. What this does is it takes people out of being the automation engines and lets things create themselves.

OK. I host the model-based testing Web page.[44] It is a very shoe-string operation; it's actually redirected to my Geocities account. So we're not exactly a mover-and-shaker. But what it does is point to a lot of resources on different kinds of models. State models are one kind, grammar models, monkey models—there [are] several different varieties. Questions?

PARTICIPANT: It sounds like the model-based testing has been a career move, a change of careers. For example, in my case, am I better off being an author or an exterminator? [*laughter, then inaudible follow-up*]

ROBINSON: I don't know; I started up as a developer but found that I enjoy finding bugs more. You know, so it may be . . .

PARTICIPANT: [*inaudible*]

ROBINSON: So part of it is: would you rather be a writer or an editor? Because, you know, the editor goes through and tells the writer, maybe you can do this, maybe you can do that.

PARTICIPANT: Can you tell me that being a model-based tester is any better than . . . [*inaudible*]

ROBINSON: I wish I could. One of the reasons I ended up in the position I'm in is partly because more people need to hear about this kind of testing. Right now, what they're doing is pulling in people from the street to test, and those aren't the people we need. We need people who understand graphs, who understand combinatorics, who can say, "I know how to actually put together a test system." And make sure that there's a distinction between activity—which anybody can do—and productivity—which is what somebody cares about.

DOYLE: One of the things that we're not talking about is the matter of testing on all variations of a questionnaire . . . We have all of these design tools, and whether we have five questions in sequence with skip

---

[44] See http://www.model-based-testing.org.

patterns or one question with five variations on wording, depending on the characteristics of the response—it could go either way. One of those is connected to this type of testing, the other is one that takes a lot of time. Would you recommend that we choose the five-question version in order to adapt your type of model-based testing? Or would you ...

ROBINSON: Well, I think that they both can be modelled. What you would end up doing is ...

DOYLE: Having to model all the many variations?

ROBINSON: ... The variation on the verbiage; can you give me an example?

DOYLE: There was a question, if you remember, on the thing Tom Piazza put out that if TCOUNT=1 you get asked this question and if greater than 1 you get asked that question? And on the screen it has the same name so it's technically the same question but the words that appear on the page are going to be different depending on whether there's an actual TCOUNT.

ROBINSON: Yes, those both sound very modelable. What you *could* do is choose one. And then if later on you decide to change it it is very easy to change the model to adapt. Because what you do is change that part and run it all, just as—I'm a tester congenitally here—somebody before gave a choice that said is your income between 500 and 1000 or between 1000 and 2000, or something ...

DOYLE: And missing something in between ...

ROBINSON: Yeah, where does 1000 go? No tester would have let you go on that ...

DOYLE: Cognitive psychologists would put that there ...

PARTICIPANT: I think this is a *very* important question about the demographics ...

ROBINSON: Oh, my God! I wasn't even here! [*laughter*]

PARTICIPANT: What do you do with your English majors?

ROBINSON: What do I do with the English majors? *I'm* a religion major.

PARTICIPANT: Do you rephrase it?

ROBINSON: No; what I'm doing is putting together a talk about how liberal arts majors make the best testers ...

PARTICIPANT: ... because they're logical?

ROBINSON: They're logical, and they delve. Remember when you used to sign up for a course because, you know, we're going to *delve* into Shakespeare or something. That's what you need—somebody who will get in there and say, "what are we actually doing?" Those are the right questions, and the right questions asked early enough can save a lot of trouble downstream.

PARTICIPANT: So you use them in a different place, the modelers from the testers?

ROBINSON: No, we actually use them all over the place; where we don't use them is at the very end of the line where, you know, we basically get young people and give them caffeine because they have to be up all night changing hardware configurations or something.

CORK: One last question and then we actually have to move on . . .

PARTICIPANT: This is basically a "checking to see that I've got it" type of question. Does model-based testing in practice consist of, for all states or for as many states as time or resources allow for, enumerating the events that can take place, enumerating the different states, and then giving that over to something like Visual Test and saying, "go for it"?

ROBINSON: Yes, and in fact I would qualify the first statement to, you model as much as you want to get testing from. So if there's an area of your application which you are sure is solid, don't worry about it. Model something where it will do you good. So, for instance, if my car keeps breaking down then I need to learn something beyond where I'm at now. But as things are now my model is sufficient.

OK, thanks.

QUALITY RIGHT FROM THE START: THE METHODOLOGY OF BUILDING TESTING INTO THE PRODUCT

Robert Smith

CORK: The next speaker that we have is Robert Smith, who's currently a visiting scholar with the Hoover Institute at Stanford. He's going to describe experiences he's had doing these sorts of testing routines at the Computer Curriculum Corporation.

SMITH: One of the, I suppose, a couple of disadvantages of speaking this late on the program is that you're likely to find out that someone has said what you had wanted to say. Perhaps better than you would have, or the conversation has gone a different direction. You also may have to modify your presentation to remove some of the more fragrant errors that have been exposed in a previous conversation. [*laughter*] So I hope that I have been able to re-thread through that just a bit.

I wanted to pick up on a couple of points. I think that incremental development, small teams, transparency, documenting first, quality throughout are really good things. Avoiding the "design chicken"—good metaphor, Jess, I used to call that "ball's in your court." I also like, however, coding first; that can be called "hacking," it can also be called "fast prototyping" if you want a nice word for it. I think that you learn a lot sometimes by, at the beginning, getting something up on the screen and letting users interact with it. And maybe that's just part of the specification process.

I like unit testing very much; I think that it helps speed up the process and brings everybody into play. I like having people who are more or less experts on certain parts of the program, although ownership—you don't want to get to the point where someone is saying, "It's *mine*," but somebody who's a specialist. Use cases, something I think Larry is going to talk about later on, a very, very important part of specification; you learn a lot when you sit down and talk to, say, your field representatives or your data analysts and so on and learn what they're going to want out of the system.

Design restraint—and I got that phrase, Pat, I think from you—is a real issue, and it cuts a lot of different ways. You can find yourself reinventing a lot of wheels unnecessarily because people come in—especially new people—and say, "I want it *this* way, I don't want to use exactly what you have," the existing framework or system. You can also find yourself stifling innovation. And I've often found myself where new people will come in and ask too *little* because they don't know the richness of what you already have. They take a kind of conservative approach, they don't ask for very much, and so you don't get the kind of product they might've. But this is a tradeoff, and I don't have an exact answer to it, except probably to try to document and explain as well as you can and get people to buy into that as much as you can, that would seem to be the best approach.

Lots of configurability within systems to allow them to change, to have things within them that are programmable. Which I gather in CASES would generally be described by putting in more features, by putting in more tags. And of course quality throughout, which is really what I'm talking about today.

What I want to do is to give a case study of work done at Computer Curriculum Corporation, to talk about particular testing challenges and solutions there, and to see if I can make the relevant CAI-to-interactive-surveys connection. And since writing this slide I've realized that CAI has two meanings here—there's "computer assisted instruction" which is where I'm coming from and there's "computer aided . . . interviewing." I almost said "interrogation;" that would have been the wrong word. [*laughter*] So we want to distinguish those two things; if I say "CAI" by accident I probably mean the education sense.

Computer Curriculum Corporation is now called, by the way, NCS Learning; it's owned by Pearsons and it's entirely different from what it was. But it was the market leader, and I think still is, in computer-assisted instruction for K-12. Founded by Pat Suppes and Richard Atkinson; I'm mentioning Pat because a number of you know Pat. Pat is actually represented here today—Ray [Ravaglia] is here as his representative from Stanford—and is still going strong. We had large comprehensive interactive courses that were intended to be used over eight, nine years, twenty minutes a day. In 1998 about four million students at 16,000 schools were using the product.

There are some testing challenges in this particular kind of course. And actually everyone wants to say that your courses are very complex. But if you look today at a lot of educational content, instructional content on the Web, and I've been looking at a lot of those things, doing some consulting for some companies—lots of things are very linear. You know, present the material, go right straight through, no change. CCC was very different because we wanted to optimize the instruction and, to do so, we were doing a lot of decision-making to decide what should be given to a particular student. Courses would contain, at the very small level, hundreds of exercise generators that would generate a large number of different exercises, and an example of that would be, "give me an exercise that has column addition and two carrots." Two clicks is a carrot. OK, so you want that kind of exercise randomly picked but still subject to that constraint through the curriculum. It's absolutely amazing how many errors can creep into a curriculum with boundary conditions like that. You wouldn't think it would be, but that presents—at a very micro level, that creates some problems, and those are probably amenable to some of the techniques Mr. McCabe suggested because they tend to use discrete little pieces of code.

But it gets worse. At a level up, we have a lot of branching things that look like the kind of thing that you would have in a survey, where you have a choice point and there are three possibilities and you could branch three different ways. We have a lot of that and it dovetails together really quickly. Then another level up from that we have something that we call "motion," which works very globally to try to match the level of the student to the level of the curriculum. And that probably sort of introduces an arbitrary GOTO statement if you want to think of this as a graph. Presenting, of course, some theoretical problems. Now, what we were trying to do was to further some optimization principles, and let me just give you a couple of examples.

One of the principles would be that if a student is having problems in a certain area but has mastered something else, let's not keep banging him over the head with the things he's already mastered. You've done *fifteen* of those perfect, you've done them perfect, we stop giving them to you. On the other hand, you're having problems with fractions; you can't multiply them. So increase the material—add to the tutorials—on the things you're having trouble with. This is one optimization principle we used.

Another optimization principle was in the area of mastery. Mastery is one of these educational ideas, and it's often defined like saying if you get 80 percent of something right then you've mastered it. [Some] use 100 percent, give or take a kid there; that's really strict mastery. But, in fact, if you think about it a little more sophisticatedly, you could, say, think of a sequence of ten exercises of a particular type. And you get the first five exercises wrong and the second five right. You attribute to that what? You attribute that you learned. And the probability of that being random is very small. So you imagine that you've learned now, so even though it's only 50 percent, you attribute mastery based on what you saw. So you can take these sequences and come up with that, and that's another of the kind of principles involved in the overall program, both at the micro level and the macro level.

And of course we had a lot of data collection and reporting facilities. I will tell you here that CCC would say to people that we are looking at four million students today and analyzing all of this data and using it to improve the courses. That was never true, unfortunately; we probably looked at less than one-half of one percent of the data. We tried to do it in a representative way. Of course, the Internet today would change that, I think. But it was logistically just too difficult.

Well, now, we did a lot of things to test this. We did a lot of code reviews, and we did lots of design reviews of a particular part. We did white box testing with rooms full of people, we did black box testing, and so on. But we also did a lot of automation, and this is just a simple kind of a characterization of the improvements that we made in it over time. A lot of it was ad hoc, not very systematic. First of all, we'd run by hand; that wasn't very satisfactory. Then we would go out and say, "let's just set this up to run, pretty blindly, all night, all weekend." And what we're really testing for there is: are there any memory leaks? Is it going to crash? Does it restart properly? All those kinds of things that are really very important but still don't get, in any sense at all, to the individual paths that people are going to take. And then, gradually, I think, we just started adding instrumentation into the product, first at the top level—where you have this global decision about what area to give each student—and then filtering that down into levels below. And so it was added in gradually, kind of ad hoc. Eventually, we came up with something called MetaDaemon, which we sort of institutionalized, and gave a name, and people could be proud of it.

Now the answer to a question that was asked earlier—Pat, I think it was a good one—you said, would we be writing and authoring and analyzing twice, or once? I think that was the question. And the answer is once. *But* we put the testing conditions in at author time. And we're Markovians, OK, so we believe that we can say, if you had a choice point and there are three choices. Well, we—lots of things would look like this. And we say that it's 80 percent here, 15 percent here, 5 percent here. And we just put weights there and code them in. Now, we could either do that at the macro level or, kind of like object-oriented programming, overwrite it at the micro level, so it could be done in a couple of different ways. We would demarcate the choice points and the likelihood parameters were stated. And, by the way, we also put in asserts—"oracles," I think you call them—but we call them asserts. If we come into a choice point, into a node, where you—for example, like before, where the choice is whether age is going to be 17 or greater—we put an assert right at the top of that. And of course you could argue that you don't need to, that the marks going into that have already tested that, but we find a lot of problems that way.

So we're actually peppering both the system itself and the actual content written in the system with the actual test conditions at the time it's written. Now, we're also Bayesians, in the sense that we're willing to alter those probabilities, if we really don't know that it's 85-15-5. So occasionally we would get in data from the field—we would get, as I said, a small percentage back—and could make changes to things. And one of the main ways we used to modify the course itself was if we saw some precipitous drop in correct answers. So you send the course out into the field, students charging along and getting 80 percent—and all of a sudden, at a certain point, the probability of success is 50 percent. Maybe something's wrong here; perhaps you jumped too quickly, you haven't introduced the material properly, you have misunderstood the level of difficulty of this material relative to the students. So that was how we would make change at that level. And we would actually put some of this—not a lot, but some—back into the testing conditions. But, certainly, in your environment, with the data richness that you're going to have, that could be done automatically.

We kept logs and scripts. Now I'm going to say I agree with Mr. Robinson; I've never had a whit of luck with these little automated test tools that, you know, save your screen, save your mouse clicks and your keystrokes and then run them back. It might be OK for regression testing, but every time we've tried to use them you find that they break right away, and usually it's because something like you added a menu item and now the mouse only moves down 3 inches when it should have been 4, and whatever doesn't work. So that's a problem. But the kind of scripts we had were not at that level; they were at this MetaDaemon level of test conditions, so they were a little more impervious to GUI changes. We also did a lot to support regression testing, and I cannot express to you my appreciation of that.

Let me just show a sample of this course; can you see? ... Global decision algorithms, and these are—maybe 15 or 20 modules here, say in math corresponding to things like addition, subtraction, fractions, geometry, time measurement, things like that. And the algorithm compares the curriculum—as a static model—to the student—as a dynamic one—and says: what should we do for this student now that would be the most optimal? And it's also probabilistic and somewhat random. But we pick that and now come down to a sequence that will just have some local branching in it. Local branching among the complexity, along the lines of what I've been hearing about branching on marriage and sex and so on. Not, probably, as complex as on something like employment, where you've got a whole lot of nodes and responses. And then an individual exercise. And the MetaDaemon information was added here, at the top, this part being done by the main people designing the course.

This part being done by the individual authors of the individual modules, expressing their sort of a priori opinions. And then the exercise, this is typically written in the underlying implementation language, C++, say, but it would be handled similarly by different people. Everybody had a role in making sure that the test conditions were built into the product up front.

Now this was mostly, I think, not very planned. It was opportunistic, it was bang-for-buck, it was "what will help us now?" We weren't really setting out to do a wonderful test product, and my guess is that if I had ever gone forward to the powers that be and said, "OK, I'm devoting two person-months, two person equivalents for the next year or two, for adding this MetaDaemon facility," I mean, how are you going to be able to sell that? How is that going to help us in our effort to sell [the product]? But, it got done, and it got done because people believed that it would be valuable.

How did we get the content providers to buy in? And there was discussion a little bit ago about that—do you do this by legislative fiat, do you tell them, "we have to?" Well, in our case, we were kind of adding this gradually enough that people bought into it because they could see the benefits, and they could see that it was going to be used immediately. And as it started to roll a little bit, you know, if you were coming along to create a new course, this was *automatically* something you were going to do because we'd kind of proven the benefits in the past. But probably the first couple of phases of it were bootlegged.

I personally like bootleg software development projects, in a somewhat large company; you have to do them and get your other work done. I think the [trick is] in getting them done in the conduct of his or her other duties. A less bureaucratic way of saying, "get your other work done while you're doing this other little thing." Lots of times some very nice stuff comes out of bootlegging, and I think it's a good way to do some things. And, as I say, it became standard.

And then some other things—these are just less interesting things. Of course, we had source control defect tracking, QA procedures designed for fast turnaround, smoke tests, and all that. Lots of unit testing and module testing, which I strongly promote.

We were missing some things, and I want to point those out to you. Customer support issues really did not find a way into our defect tracking database. And, actually, this happens in a lot of companies. So, the customer support organization was dealing with the customers, and they were getting, you know, all these issues. And many of them were, you know, "I didn't plug my terminal in" kind of things. But there should have been some more automatic way to get them into the engineers' tracking database. We had a political problem there; it ended up being solved by a committee meeting every couple of weeks, and they would negotiate which ones would be migrated, but that's not the way to do it. And I think the QA process can never be integrated enough; you can never build it in enough.

Now, I wanted to compare, here, CAI to CAI, computer-assisted instruction to computer-assisted interviewing. And I think that there are a lot of points of comparison and also some differences. You're looking for questions and answers, really, in the educational context, and you generally have an idea of what's the right or wrong answer when you ask a question, as contrasted with what you're doing—a tutorial or something of that sort. In a survey, it's more likely to be a neutral response; there may not be a right or wrong answer. There may be an inconsistent answer, but there might not be a right or wrong one. I don't know how much difference that really makes; it certainly makes a difference in how you interpret things. And from a QA perspective, I'm not sure.

Here's a big difference. We were taking, of course, wrong answers and providing error analysis and tutorials. And, in some cases, back-ups. And we would back up by taking the student back into part of the course; he's not doing well in fractions, so let's take him out of fractions and move him back to a more elementary point for tutorials. And that back-up generally left all of the data on performance that had been done later, because it was still considered relevant for performance data even if it said it wasn't enough, that he wasn't performing well enough. But, in the case here, when you back up I think you have a lot more issues of what you do or do not do with that data, and I don't understand it all, but it's very interesting.

We were branching for optimization; we were branching to try to curtail the amount of extraneous information given to a student relative to their needs. And I think you're doing something different: you're branching for relevance. They're something similar, in effect, but there's a difference in purpose, looking for relevance; it's not relevant to get spouse information from a single person.

Here, I speculate: our courses, at least, changed slowly. The course that CCC is shipping today—or that NCS Learning is shipping today—in math is very similar to what it was ten years ago. And your surveys may change more quickly; I don't know. I heard you saying 40-year-old questions are in there, Pat, so maybe that's something, too.

Now here is a really critical question or a point; you may be able to exploit this. We had a large audience of students and, mostly at the time, were shipping on CDs. And, so, when you go gold on that and want to ship it out to 16,000 schools, that's a big job and you want it to be right. You have, as I understand it, a fewer number of FRs, and people who are going to be involved in the test administration or the survey administration. And that may be some difference, I don't know; you might be able to try some more experimental things sometimes, but we can't; it's got to work.

One difference here is the standard of quality that has to be attained. OK, I wrote that when I was asleep, so I wa speculating.

What I would suggest is that you see if you could do this, and start with the CASES product—just as a methodological suggestion—and add some statements to that language that would allow you to say, at each choice point and each error/consistency point and back-up condition, what you're expecting. And see if you can build some automated testing around that, if you haven't already done that. And if you've already done that, then I'm apologizing. And I also put a quote here from a former QA manager of mine—and, Mr. Robinson, this is your desire—he said, "people know where things are going to fail." And he would select the particular developer who came in late, and looked like he'd been run over by a truck, and stayed late—and that's the person whose code would get tested most. And, boy, is that a winning strategy. So try that by all means. Of course, you may not have any; we're from *California*. [*laughter*]

DOYLE: And we're from Suitland … [*laughter*]

SMITH: This is just module testing. Test during development, and so on. Again, I like that approach very much. Test on module integration; acceptance testing for each build or revision, including … I like to be able to have a build at 8:00 in the morning and know by 9:00 whether that build is going to be testable. I mean, you've got to be able to run it through its paces enough so that, if it isn't, you can tell the developers to redo this and get it back to us; it failed. You don't want to spend all day waiting on that. And, of course, I've already mentioned regression testing. In particular, the suggestion I have here is authoring into your survey some of the conditions by which you want it to be tested, and automating that process.

And these are just a couple more issues, leading into some of Larry's talk that's going to be coming up shortly. We did coordinate slides, so that we would not duplicate *too* much. You might pursue asking yourself—if you already haven't—how long would it take to make a new survey? What would you like that process to look like? And I'm talking here about the requirements of the process. And to put as a goal for yourself what used to be called "Internet time;" a couple of years ago, during the dot-com boom, everybody talked about "Internet time" as being very fast. And people aren't talking that way anymore, but I think there's some benefit in doing that. And I heard earlier some people suggesting to you that there might be some benefit in terms of your competition with other survey organizations, or quasi-competition.

I had as a suggestion to build an interactive GUI environment for development, and I'm pleased that Mr. Bethlehem had a good deal to say about that earlier. I believe that he was—I think it was his, there were a lot of talks so I might be confused. I would give you a possible suggestion there; I would say that something that was XML-based, for example, but that could output CASES as its object output so that you could use all your same deployment systems, but would show you various views including texts, graphs, flowcharts, and so on, and would impose the authoring environment—impose conditions on the authoring process. So, for example, you require that there be test conditions. You could ask it: what am I missing here? What do I need to fill in and complete? And it's a bit easier to use in an GUI environment than, say, in a scripting language, which is what I understand CASES to be to date. But I don't know CASES, so we might talk to Tom about that. I think that some of the work discussed here earlier on the documentation side could really be flipped around and made into that kind of a system.

And then, of course, there are some modern methodologies such as UML and extreme programming, and an upcoming talk will deal with that—Larry's talk.[45] And I think there are a lot of good ideas here, but I do suggest that you adapt and evolve things rather than follow blindly. I'm not very ideologically based myself, so: use what works, and find that many things can be a part of success. Any questions?

BANKS: I very much like what you're saying. As a statistician, I certainly support using the design of experiments to find new ways to develop surveys. It seems to me that the same problems arise in software

---

[45]UML, the "Unified Modeling Language," is a language intended for the modeling of complex computer systems. Specifically, it is meant to provide a mechanism for documenting and visualizing object-oriented software systems.

performance testing; there are strategies that they use that have been put out. [Has] anybody done any studies of relative efficiency of code? Say you're trying to do something, like linear programming, and you have two things that do it: one uses one style of performance testing and the other uses a different style. And then after each group is finished the question is, what is the difference in the testing?

SMITH: That's a good question. I don't know enough of the general literature in testing to say that I have an answer to that. I think a lot of these things become a question of how built into the process it is. And so, to me, the idea of taking any form of testing methodology *whatsoever* and sort of plugging it in at the end is . . .

BANKS: Yes, but I don't mean plugging in at the end. What I'm asking is, say Team A integrates one philosophy in their work. A second team, Team B, uses a different strategy, over Team [A], and at the end of the day the question . . .

SMITH: Controlling the other underlying variables is the question in that . . . I don't know the answer to that. Maybe Tom does? I've gotten the name wrong; I'm sorry, your name is?

ROBINSON: Harry.

SMITH: It was either Tom, Dick, or Harry. [*laughter*]

ROBINSON: We've tried similar things to what you're describing, and the problem is that it's hard to give the same job to two different teams and justify that. What we've had to do is to give the job to one team, and then have them do it again in another way . . .

McCABE: As another Tom, Dick, or Harry at this conference . . . I'm not a statistician, so tell me when to quit if you've seen all this, but there's a literature about error seeping that says you can intentionally, in a piece of software, put an error in. It doesn't really pertain to software, where we have plenty of errors [*laughter*], but the place I've seen it apply is when an organization puts the product out in, maybe, four or five sites simultaneously. While testing the product, you *know* where that error is; you know where some null errors are. Then you match the weights from the field sites to the known errors. And you typically have different characteristics in terms of the kind of testing, black box, white box, and whatever. And from that you can make some inference . . .

SMITH: I think that's a good point, and it reminds me that when we had the new product in beta release, we were kind of doing a model test, selection, for the people to whom we sent the beta product. We would pick a school where we thought they had very good tutors, very good proctors. We would pick one that we thought was a little sloppy. We would pick a school that was primarily using it in math and reading, we would pick one that . . . So, you know, we would try to have some sort of balance regarding what we thought our overall population looked like, for providing the product in new release. I think that's different from what you're saying, but it reminded me of that. And, as I understand, how you would be dealing with that would be kind of cut the development right here, it stops, and send to external organizations using different approaches.

McCABE: [*very faint on recording*] One of the problems with this, it turns out, with universities is that it's *way* too expensive to use as these operational sites. . . . But an operational company will usually be able to field a couple of operational sites, and typically those focus on different priorities. Or they pick one site that is known to get something done quickly; that information is relevant. And then you get some comparisons among the methodologies.

GROVES: I see that as an interesting comment because it would be great to have something that allows us to judge which of those sites. The question is the criterion . . . [*trails off to inaudible*] I wonder what kind of evidence you can produce to say: I prefer this method to that one? What do you fit into the criterion?

McCABE: I think it's really important to realize that there's no single testing method used at any of these places. [*inaudible until closing comments*] I think that the place that test well do so because they have at least four or different methods that they use at various stages. I don't know of any place that has just one thing going in testing.

SMITH: I think that a lot of those methods are very good, and I've also found that if you integrate them in a regular way, then they smooth out and the problems with them start to diminish. For example, code review can be a threat to people who have never done it before, because they remember back in college when their work was being graded or something. And after it becomes commonplace and everybody's had that chance to go around the table, it becomes a very different thing. And you find a lot of problems that way, and you understand the code, and you do some cross-training to some extent. And I think that everything else you mentioned—unit testing, regression testing—if regularized does kind of smooth out. And costs become less. Would you concur?

McCABE: One thing we didn't talk about that we ought to comment on is relative cost of testing software versus surveys. In software, it's quite high; in software, it's often, in the total budget, maybe 60 percent. Now places like Microsoft, where you're shipping millions of products, the testing cost tends to be very, very high because the cost of errors is so high. Versus more of an engineering shop. But, still, we're in the software business and producing, and so the costs of testing can go up to 60 percent. Now how does that compare with surveys?

DOYLE: It's a small portion of the current budget. Of the total budget.

SMITH: Mr. Robinson—Harry Robinson, that is—a question about Microsoft ... has the advent of the Internet with the ability to update modules sort of quickly with patch packages and so on, has that in any way changed their feelings about testing? Do they have any sort of a sense of, "well, if it's wrong, we'll just update it later?" Or ...

ROBINSON: There might some of that, but now that the Internet has opened the door to so many security bugs, it's actually swung the *other* way.

SMITH: I always hated having to buy a new computer and load four packages on it—I mean, a brand new computer, you know. But ... any other.

CORK: That's a good breaking point, so let's thank Bob.

INTERACTIVE SURVEY DEVELOPMENT: AN INTEGRATED VIEW

Lawrence Markosian

CORK: The next speaker is Larry Markosian. He is currently a technology transfer consultant at the NASA Ames Research Center. He founded Reasoning, Inc., and was for several years the product manager for their principal product, which was a tool for automated software defect detection.

MARKOSIAN: OK, so, my talk is going to expand on some of the themes that have been discussed thus far. And I'm going to perhaps suggest some solutions that are a little more specific to addressing the problem that we've seen. So, the roadmap of the talk is to review some of the basic challenges. Then we'll look at some testing principles and a couple of specifics—a methodology for programming that Bob [Smith] alluded to in his talk earlier—and also a set of integrated tools that actually will work well with this paradigm. And then we'll just do a quick summary.

So the challenges include many of the same things we see in other software development projects. In fact, there is relatively little that I've heard thus far that uniquely characterizes this problem. One issue that has come up time and time again is the large state space, so we'll look at ways and tools that can help address that. But, in fact, the reality is that the large applications put out today also deal with a large state space, for the reasons that Tom McCabe pointed out earlier.

I've also heard that there's little time for testing, and little formal testing being done currently—you said that it was a very small portion of the budget. But we've also heard that there are bugs in the products that are being delivered to the customer, and these include things such as unexpected behavior—what questions are being asked, and occasionally crashes in the system that probably upset the FR quite a bit. Another problem that I've heard is the relatively small amount of use of development artifacts, software artifacts. And that doesn't mean only pieces of code but also the tests—whatever the test suites are, along with the documentation. And then, finally, we've heard that there's very little time to produce documentation.

So all of these problems are common in many software projects, and there are a couple of principles that we might pursue in finding a solution to these problems. They are also becoming more widespread generally in software development. One of them is to integrate the process—the development process itself. That is, how people actually go about developing software. And the concepts here are that we want to have close involvement at all stages of the lifecycle by all the stakeholders in the process (or at least in the product). And perhaps that bias comes from my own previous position, working as a product manager, where I had to be sure that all the stakeholders were in agreement from the very beginning. But actually we can do a lot more here than what I was able to do.

Another concept that's important here is iterative development, and a number of people have mentioned this before. We don't want to have—particularly if there's a hard deadline—everybody spending a lot of time initially doing requirements analysis and then, you know, design and implementation. And then, at the end, we find that we send it over the fence and the customer rejects it because of a requirements issue that could have been found much earlier in the process. So, we should have iterative development, with small iterations where each iteration produces a useable—or at least a testable—product. And that concept has come up here before.

Well, what we haven't discussed is a particular methodology that incorporates these concepts. And there are others as well; I don't mean to be advocating this particular one. This is a an example, and you should look at its strengths and its weaknesses before you get into it.

The other principle is integrated tools. We should try—the two ideas here are first, that, during the lifecycle we produce many software artifacts. Requirements, specifications, test cases, code, design, and so on. We should formally try to capture them as much as possible. And then what we can do is use automated reasoning techniques to derive the next level from the previous level—[to] help go from the specification to the implementation, for example.

So, let's take a little bit of time to look at the interesting aspects of "extreme programming" (XP). This is a style of development that's particularly suited to program domains with changing requirements. And what I've heard is that this—that CAI is one of those domains, that we could be going along on our goal of shipping the questionnaire and then, at the last minute or at various points along the way, a new requirement will come up. "Congress just passed a law," I heard. So this has to be reflected in the code. But, more generally, even without that kind of last minute requirement, there's the problem of having the customer understand the requirements that they are specifying. And usually the customer has not thought them through well enough to realize that there are significant implications that are unaddressed in the requirements. So by doing iterations—rapid iterations—on this and continuing to do requirements analysis through the project, we can reduce that risk.

OK, well, it says here that this technique is particularly appropriate for projects with high risk. We've already identified some of the sources of risk here. And one of the problems here—one of the great sources of risk—is that if you have to deliver by a certain date, ship at a certain date, then the software project is automatically at a high state of risk.

So extreme programming is also useful with relatively small development teams. There are cases when it's been applied successfully with larger development teams of 30 or 40 people, as well. And if you do have larger teams than that, then you might think about breaking the project up into several smaller subprojects in a rational way.

One of the requirements for doing extreme programming is a commitment by all the kinds of people involved to work together shoulder-to-shoulder. So, we're not going to have developers sitting there doing development *on their own*. We're going to have managers involved and going to have customers involved—very closely involved—with the developers at every stage of the development process. Another requirement is testability because—as you know, or at least will see in a little bit—the mantra is, "test early and test often." If we can't, if we have a project that's not readily testable or an application that's not readily testable, then we're not going to be able to do this.

So, here's one of the stages in extreme programming. The first stage is usually the planning stage. And, in this, user stories are written. A user story is like a scenario, but in the context of extreme programming we try to keep that very short. Just several sentences describing a use case. So it's less formal and is shorter than a standard use case. And, again, the concept is to get everyone on the same page very early in the project. Then, there's a release planning session that creates the schedule, and again the idea is frequent, small, testable releases. The project—in order to achieve this—is divided up into iterations. Iteration planning starts each iteration, and there are techniques for measuring what's called "project velocity." Again, the concept is that you're going to have releases very frequently, like every three weeks; on a small project, maybe even more often. So that we can very quickly monitor the status of the project and determine whether we are meeting our goals.

One other important part is that we fix the process, we fix [it] when it breaks. And there are various ways in which it can break, and we'll talk about that a little further.

The stand-up meeting ... the purpose of the stand-up meeting—and that's a little bit interesting—is that each day starts with a stand-up meeting where all the participants get together. And the purpose of the stand-up meeting is to avoid these *long* meetings, these three-hour meetings where people sit around—they're very low bandwidth meetings, people fall asleep, and very little gets communicated. What we do at the beginning is have a very short meeting, with a time limit set, where all the issues are brought up, and then the development group breaks up into the natural teams to address these issues. And how that happens I'll discuss in a bit.

The next step is generally design, and the guideline here is to choose a system metaphor. A system metaphor gives you a way of establishing a consistent communication style and a consistent communication language, all the way down the requirements, down into the specifications and the code. So that people have a similar set of concepts—the class names are consistent, the levels are named consistently, and so on.

Another concept is that CRC cards are used for design sessions. If you're doing object-oriented programming, these are class-relationship-collaborator cards. These describe, for each class, what the methods are and what the other objects are that are involved in that. So the idea is to quickly get to an outline that's understood and mutually agreed upon.

Another principle is that no functionality is added early, and the reason for this is that we don't want to get ahead of ourselves. This is a project plan and, as we'll see in a little bit, a key concept is that most if not all of the implementers on the project are capable of implementing any piece of the project. So if subteam A gets done quickly, they don't go off and implement something for the next mini-release. What they do is go off and implement a piece that may be running behind.

Then another mantra in XP is to refactor whenever and wherever possible, in order to eliminate redundancy and so on.

Now we get to the point where we're actually coding. The customer needs to be always available. The customer is dictating the requirements of the project, and the customer has to be involved—particularly involved—whenever something, a piece of it or an iteration of it, is done, and there's a need to evaluate it. Because that's what's going to help them—actual pieces of operating code, running code, even if it's very quick and a small part of the overall system that's being developed. That's what's going to get the customer thinking about the requirements further. And you'll also find out whether you've done the right thing. Maybe there were just some omissions.

Code needs to be written to agreed-upon standards. And, again, the idea here is to support the notion that we'll get to a little bit further, which is that the whole code needs to be understandable to the entire team. We can say—it may very well be that there are some parts that are only really understandable to the expert who wrote them. But, to the degree that happens, we increase our project risk, if something happens to that person. So we want to have code written to agreed-on standards, because that helps communicate things to the group and it's easier to move people around.

We also want to unit test first, and there's been some talk about that thus far, too.

Production code is pair-programmed, and what that means is that instead of having two workstations with one programmer at each we have one workstation with two programmers working at that workstation, on the same code. And they shift, so that they are each driving at different times. The evidence is—at least in the XP world—that this really works, that you get much higher programmer productivity by having two people working on the same code at the same time. In my . . . in the projects we did at Reasoning, we didn't actually do that, so I can't speak from experience on that. Most often, and in my case in particular, we used a number of the reductive, a number of the principles, but not the whole dogma of XP.

Integrate often—that goes pretty much without saying, because we're going to have many releases. And at each point we're going—as part of each release, we will want to integrate, and usually integrate prior to the release. Now there's a concept, also, of collective code ownership, and I've alluded to some of the components of that; we'll take a look at that in a moment.

There's also the concept of no overtime, and that certainly didn't work for us. But the notion is that if you have a good project plan, well, the fact that you're requiring people to put in overtime on that means that there's something wrong with the plan, that was not planned in. And so we should really be trying to get the project right—maybe not completely at the beginning, but certainly incrementally as you go along, refine it to the project plan.

Here is a slide I promised on collective code ownership, and notice at the top here it says, "Move People Around." And in the center it says, "Pair Programming." So these are two of the key concepts—we want people, the developers, to adopt any role in the project, work on any component of the project. And pair programming I've already indicated what's involved there. Now certainly there are going to be cases where you have applications running over a network and there's going to be some network expertise that's required, and we're not going to be able to move people around that well. But, to the degree that we can, we reduce risk because everybody understands—or is fairly quickly able to understand—most aspects of the project.

And finally we get to testing—not "finally" in terms of the project, but "finally" in the sequence of slides on XP. All programs need to have unit tests; the code needs to pass the unit test before it's released. And I guess that the most important point here, really, is that acceptance tests are run often, almost continuously run, and the scores published. And the reason for that is so that everybody can—it's part of this shared responsibility for the project, and everybody needs to know where the project stands.

Extreme programming seems to be coming into a lot of popularity these days, and there's a Website for you go to go to learn more about it. And I want to emphasize, again, two things. First of all, it's not clear that you need ... you don't *need* to do this. The methodology suggested here is not the only methodology that could be useful in developing questionnaires, but it's one that I think—from what I've heard—is going to help reduce risk an enormous amount. And even just picking out some of the basic principles from this paradigm will help.

OK, so that's my perspective on the integrated development process; now, let's take a look at some integrated toolsets that can help with some aspects of the development, with many aspects of the development. OK, first of all, what we want to do using integrated tools is first capture, formal capture, of logical artifacts. Now, I debated for a while on this slide whether to say "machine capture" or "formal capture;" I felt that "machine capture" was too weak and "formal capture" was too strong. But, that's where it ended up. By "machine capture," we might simply mean to make sure that all documents are available in a collaborative development system to everybody on the project. But that's a little bit weak because it won't really allow the use of tools to use on these development artifacts. So we want them captured in a way that supports the use of tools. And the current standard—for capturing requirements in design, at least—is UML. There are extensions of UML as well, but let's say that that's one standard. Historically, there have been many others that led up to this, and they are still in use—OOE and OOA are others. Once the designs and other artifacts have been captured, then you can apply tools to manipulate them. And there are a variety of tools that are available, so let's look at what some of them do.

The common tool capabilities are to capture the requirements and then to trace those requirements through the development lifecycle. And we'll mention some of the tools a bit further down. Other tools are available to model the usage scenarios and use cases—that is, to model usage cases to help you understand usage scenarios. There are also tools for migrating from use cases to UML sequence and collaboration diagrams. So, again, these are tools so that if you capture your designs and models in a formal method then you can move further along the development lifecycle. We also have tools that help you build class diagrams and actually generate code from the class diagrams. Now, the code is generally of the form of a skeleton but it help move things along. But there are also cases—particularly in the case of finite state machine operations—where you can actually generate a lot of the code. CAI seems to be particularly amenable to finite state modelling, so there are—within UML—tools that allow you to model state machines using state diagrams. And then there are ways of modelling component relationships (the components of your software systems are the source code packages and their relations), and then your delivery system and the components of the delivery system as well.

There are various tools and toolsets that are available, integrated—more or less integrated. Rational Rose is probably the most widely used integrated toolset for this, and then there are other products from these companies. And I mention this one because it's open source; I don't know how appealing that is to you, whether you're into the open source community. And then there are other standard toolsets as well.

Let's take a look at some examples of use cases. Now, this is the beginning of a use case to look at use cases. Here, we've got defined a use case, and then we've defined the other classes that are related to it and what the relationships are between this use case and the other classes. So we've got use cases, and use cases are related to users who express them; they are related to analysts who understand them, or analyze them, or try to understand them or complete them. And then there are designers down here, operating on them as well. And, of course, programmers and testers out there; I don't know why I chose that, it must have been late at night. So I chose that guy ... So here's an example of a UML use diagram. Now, I'm going to be showing you only three diagrams from UML, but there are twelve in UML 1.3 that cover a lot of development/design activities. So here's a case where we have a sequence diagram; these are the classes. We have a caller, a phone, and a recipient. Now this is not intended to model anything you're doing in CAI ... So, in the sequence diagram, we show the messages that occur, that are sent from one of the objects in the diagram to another. So, a caller picks up the phone, the phone replies with a dial tone, the caller then dials, and the recipient replies with a ring notification, and then finally the phone picks up. Actually, this should probably extend through ... Yeah, the recipient picks up the phone and says, "Hello?" Now, UML supports defining these with a temporal relationship on them. So this is an example of a sequence diagram.

Now, finally, what might be even more relevant in this case [are] state charts. There's been a lot of discussion of state machines, and the state charts are a way of representing state machines. They have several advantages over simply drawing the complete finite state machine, which may—in fact—have too many states to even be drawn. First, they do allow you to specify states. They also allow you to abstract the states, and the key—I think—to getting control of the state explosion problem, is abstracting the states. So here we have a state, state B, which has two substates, B_1 and B_2, and here's a state C that has two substates, C_1 and C_2. And there's another state, A, that has no substates. Another thing that they allow you to do is to aggregate the states, so that—the concept here is that once you move into this state you're actually in both of these states at the same time. So there's a bit more expressiveness here than you find in a traditional state machine model.

Now, I have a couple of notes, and I want to say a few words about how the models like this differ from the questionnaire. I think you asked a question about how, whether there will be a duplication of effort and how much there will be, and there was some discussion about this. I guess that my perspective on that would be that, to a large extent or significant degree, if you get a state machine model correct, then you're very close to getting the code correct. There are actually tools out there that will allow you to automatically generate the code, or a lot of the code, from this state machine. Also, you'll be able to get test cases generated, with some degree of automation, from the state machine and the state charts.

I guess the other point is that the states—when you're doing modelling of a questionnaire, you can model them at various levels of abstraction. This abstraction can be done at various levels, so you can begin testing—as Harry mentioned—early. And then as other features become more interesting—for instance, I think you brought up the case, or someone brought up the case, of five questions or five variants of the same question reflecting different outside factors—well, it's really one question but it's being asked in five different ways. So we don't need to model every one of those; we don't need a state for every one of those possible questions. We can abstract all of those into one state, and then we might—so at a certain level we'll be doing testing to make sure that we get to that question, that abstract question. And later on as we refine the model further we would want to know whether we asked the right variant of the question, and so will have some more states to deal with. But they'll be abstracted so we can do the testing at the one level, and when things are further along we can do further refined testing.

Now, in addition to the UML tools, there are various non-UML tools that are, to a greater or lesser extent, integrated. These are all from Rational. I mention Rational as one example because it's something of an industry standard, but I have no commitment to that and I don't advocate that particular tool on this project. There are others—for instance, ILogics provides state chart tools and other UML tools. And they usually provide a very similar range of coverage of the UML. So, non-UML tools that will help testing are, for example: Purify, which will detect memory leaks and other structural bugs; Quality Architect, which again is a tool from UML that automates test case generation and management (I spoke about test case generation earlier); and then various other tools. There are alternatives to all of these particular products from Rational.

In summary, first of all, we should consider a different model of the development process—one that imports some, if not all, of the principles of XP in order to reduce risk. Another problem that we should address is the loss of information, or doing a lot of work to get from one step from another. And that can be automated; when the software artifacts are formally captured, then we can apply tools and reduce the level of effort that's required there. We also gain a greater degree of traceability from the implementation back to the requirements. XP gives you continuous monitoring of the project state. And I don't mean status reports by this; during the stand-up meetings, you're basically continuous monitoring so that everyone knows what all the pieces are. This is quite different from drawing a Microsoft Project document and then trying to plot where you are along the way to meeting milestones. Collective ownership enhances understanding among participants, and automated generation of software artifacts, again, imports this notion of risk reduction by allowing the project history to be captured and preventing loss of information.

What was not really in my talk but responding to some of the issues that came up, everybody seems to have commented on the cost of fielded bugs. So, [*laughter*] I have to do that, too. This goes *way* up, if I could have just one more minute? OK, the current organization that I work for is NASA, and we've had some *spectacular* bugs recently. [*laughter*] So the thing to keep in mind is that, yeah, there's a cost involved in fixing bugs that I think has been widely estimated as too low, here. For example, if you go and look at Capers Jones' documentation here, the ratio between the cost of finding a bug during the development, when the programmer is able to come in and look at the results of the unit tests on the things done the day before, and the cost of fixing a fielded error is on the order of at least 1,000.[46] So there's an incredible difference, and we've seen that in our customers as well. Well, in Mars missions, there's another

---

[46]Capers Jones is chief scientist emeritus at Software Productivity Research, Inc.

cost, and that's collateral damage to the business. And I don't know what that cost is here; usually it's an intangible cost. But you probably have some sense of the level of frustration and so on, not knowing that no vehicles are going to crash or that nothing is going to sink the project.

ROBINSON: So I guess it would be safe to say that your costs are astronomical. [*laughter*]

MARKOSIAN: The execution costs, yes ... Any questions?

DOYLE: One of the things that we had done was to have people specialize in a particular survey, and often you end up with one programmer on a complicated survey. But a lot of these testing things might suggest that we need teams; we need more than one on a project. Are you recommending that we change our staffing to that, instead of having person 1 on one project and person 2 on another, we have two people on the same projects, or something else?

MARKOSIAN: Well, I think ... I can't really answer that because I don't know enough about the application. In the applications I've been describing here there have been multiple people working on one project ...

DOYLE: And the minimum number you've listed is two programmers ...

MARKOSIAN: Right, right. Well, one would expect, then, this has come up before, that there's a lot of experience that should be shared among the developers, even on different projects. And that's what I don't understand well enough, to advocate a position there.

SMITH: I risk making things too light here, but there's a saying that the *optimal* number of programmers on a project is two. [*laughter*] For example, Unix was originally built by two people at Bell Labs, and all the Bell executives said that they didn't know it was happening ... and if they had known, they would have stopped it.

MARKOSIAN: Harry?

ROBINSON: For the productivity of the pair programming, where do you measure their productivity? Is it the amount of code that actually makes it to release-level quality?

MARKOSIAN: No, their effectiveness is in producing code of acceptable quality ... well, yes, it's ... certainly, the metrics that are used such as lines of code—and we could look at those, function points is another metric—all have their problems. In the area of XP, I think that the focus is on quality and in meeting deadlines, and in reducing risk, so the best metrics for that, I don't know. Any other questions? Okay, thank you.

PRACTITIONER NEEDS AND REACTIONS TO COMPUTER SCIENCE APPROACHES

Mark Pierzchala

CORK: And, finally, our capstone presentation for the day—and the capstone for the testing section of the workshop—is by Mark Pierzchala, who is a senior systems analyst at Westat. Previously, he was a mathematical statistician at the National Agricultural Statistics Service. Mark?

PIERZCHALA: I just want to make one comment, first, on the presentation by Thomas McCabe. Remember that slide he showed where there were four ways you could mess up code—like jumping out of a loop and so forth? That's often what's specified that I have to do ... [*laughter*]

I was invited to do this talk, and I got some papers off the Internet, including some by Harry, there. I went through and tried to apply it to my experience in producing computer-assisted interviewing instruments. And, so I'm just going to go through these fairly quickly, but Pat and others have already given 90 percent of my talk. But some of the things we test for in computer-assisted interviewing—certainly, valid values and flow, but also the hard edits, the soft edits and the computations. But then I have a whole second slide, and there's a reason I have a first slide and a second slide. But all I want to say is the stuff on the second slide— I'm not going to enumerate here—this is the fuzzier stuff [*See Table II-1*]. Things like usability and that kind of stuff, or getting the question text right when your question is all fills. That, to me, is a bit fuzzier. But I think that the point of this slide is that, often, when we arrive at a question, we want the tester not to merely verify that we've wound up in the right place but to verify 25 or 30 things, all at the same time. Or maybe they'll do it in phases. But there's a lot of stuff going on.

A lot of the challenges that we meet have already been enumerated; I have a few ones here that are not enumerated. But let me just go through. We talked a little bit about scale, but I have an example— the Bladder Cancer Survey, where we ... interviewed cancerous Spanish bladders [*laughter*]. This was actually conducted in Spain. I could have printed out the paper questionnaire, and I was going to just to be able to take the photograph, but then I figured out how many hundreds of dollars it would have cost and how many trees it would have killed to do it. So I never really printed it out. But it's 16,000 pages in the questionnaire, and it's in two languages, so it's taller than a person if you were actually to print it out. And Westat did this survey all on a laptop. And I'll say—and you can look it up, since this whole presentation is in your book—it actually worked pretty well. I'm not saying there weren't

Table II-1 Areas of Testing Within
Computer-Assisted Survey Instruments

First Slide:

- Valid values

- Flow

- Hard edits

- Soft edits

- Computations

Second Slide:

- Screen appearance

- Proper question and edit text

    - Static text
    - Dynamic text fills done properly
        * Pronouns, possessives, adjectives, etc.
        * In multiple languages

- Sample selection

- Multimedia display and play

- Data are correct

- Navigation

    - Non-linear navigation
    - Ad hoc navigation

- Pop-up help screens

- Interviewer understanding and usability

NOTES: Items from the first slide are more mechanical in nature and more readily suited to automated testing, while items from the second slide are context- and interface-specific features that may require human interaction for testing.

SOURCE: Workshop presentation by Mark Pierzchala.

any bugs, but I'm very pleased with this. But this gives you an idea of scale we sometimes go through.

And then something I haven't heard mentioned yet: versions of questionnaires. National Agricultural Statistics Service has offices in virtually every state and every questionnaire for the quarterly Ag survey in every other state is different. In NASS, they actually generate instruments. I mean, they have a spec database and they generate the instrument; there's no programming after that generation. And that is what I call the Impossible CAI Program, and the only way to solve that was actually to generate the questionnaires. About half the questionnaires are CATI; half are on paper with field interviewers; and interactive editing on both modes of data collection. And that works. But that's just another challenge that we have. Over a thousand production instruments have been produced so far; it's been going on for seven or eight years. That's another challenge.

Then, we had these longitudinal surveys. We've gone over that before, but we will often visit the same person—especially in agriculture, you'll sometimes visit the same farmer, you know, 20 or 30 years running sometimes. And it's all dependent interviewing. How does the CAI industry test? Well, we test from specifications, and that's manual testing. And we have some scripted testing, but that's still manual. We have some ad hoc and targeted testing, and that's manual; what I mean is that somebody is pounding away on the keyboard and then we do it over and over and over again. And I don't think that's so rare in the computer industry. But I will say that [there have been] some recent advances in our industry. There is now more formal version control and build procedures; we have pop-up GUI error reporting dialogs from within our instruments, we're using tracking databases. And we're getting better at it. I'll say the last bullet here—I think that people are finally starting to catch on, what it really takes to test. Not always the case, but it's starting to catch on.

There has been some automated regression testing. Westat, where I work, has experimented with WinRunner. RTI has a WinBatch playback utility. But the question that always pops back to us when we have these automated script-playing systems is: when do you implement it? How do you update a script when the specification changes? And what is the overhead to build scripts? So that's sort of the down side to automated script testing, as far as I can see it, anyway.

I will say that one thing that I don't think has been mentioned yet is that sometimes the specs are wrong. They're inconsistent. And it doesn't matter if you've got a database or not; you can have one line where this Access database says, "do A," and the second line says, "do B," and these can be inconsistent. So we often have updated specifications. And sometimes, of course, us programmers get it wrong and we . . . you know, I will admit that, and I've done it recently myself. You can read the spec wrong. So there's a lot of iteration and rework.

My reading of model-based testing is that test scripts are generated automatically based on specifications. And one of the things that model-based testing tries to do is to keep the number of scripts reasonable. And then there's something called a state table; we've heard a lot about state tables but, basically, here's your state you begin at, here's your action, and here's the expected result. The state table can be hundreds, thousands of lines deep. Then you run the scripts in an automated system and you analyze the results. This is what I got from reading four or five articles on model-based testing.

I'll say there are some advantages from my standpoint. You don't have to hand-record the scripts. Therefore, you can execute many more scripts. Also, some scripts will exercise the application in ways that hand-written ones will not, and I think that's very positive. And since the scripts are generated you can apply the tools much sooner, and you can overcome some of these robustness questions—for instance, what happens if you delete a question or insert a question. Scripts are not targeted, and that's not necessarily a bad thing. But I will say that you're probably going to want to have some targeted scripts, too. And one of the things I like is that you don't have to test all combinations of valid values, and I have some examples here.

Now, where [might] model-based testing help? Let's just take the valid values as being OK, and from that you can generate scripts. And one of the things, when you read the literature a bit, you keep hearing about constraints—you know, you have these valid values but then there are constraints. And, to me, the flow—I mean, the skip patterns—are a way of showing constraints. So you have scripts, you subject them to constraints, and then you can test some other things—you know, whether the hard edits pop appropriately or not, and so on.

I have a simple example; I like Harry's diagrams better, but these will get the point across. And I actually programmed this in Blaise; it's a four-question instrument. But, just say that—I did a very categorical example where you have 2 categories, 21 categories, 2 categories, and 21 categories. And there are essentially two paths through this, and these little symbols—these are edits, but I won't being going into those today. My perception of how model-based testing might be executed on such a simple example is, first, I would test the flow. Then, using the flow as constraints, I would test the hard edits. Using that as constraints, I would then test the soft edits, and so on.

I'm trying to test intelligently; I'm trying to cut down on scripts. Because—as you're going to see—even with such a small example there can be an enormous number of combinations of values. For instance, on the example I gave you, the number of possible combinations of question values is 1,764. The number of pairwise combinations is 613, and the reason I picked up on this is because I read it in the literature—it says that you don't have to test all combinations, maybe if you test pairwise combinations you might be able to be as effective, and that's *certainly* going reduce the number of scripts, for example. But we've seen that there are only three paths through the questionnaire. So how might we be more efficient about that?

Perhaps we look only at the regions; they come from the diamonds in the flow graph, the decision points. For example, I might test the region boundaries. And then maybe I'll just pick a point in the middle of these boundaries. And perhaps I can test effectively, to my comfort level, using just 12 scripts rather than, say, 613 scripts. This is the way I picked up from the literature how you have to start looking at this stuff.

So if flow testing is OK then look at the hard edit testing, and the idea is to cut down on the number of tests using the flow as the constraints. And remember the number of pairwise combinations was 613. But if I use the flow as constraints, then there are really only 104 combinations that are valid for some of this stuff. Now, I will say that—in the articles I read—there are these algorithms, and they were mentioned earlier: the Chinese postman's algorithm is a way of going through the state table with as few scripts as possible but still cover all the necessary actions. I think that this is a beautiful idea; I wish I knew how to do it.

I would say that, because of the advantages of model-based testing, it is worth investigation. Implementation will be a challenge because there are details to work out. I'd say that I read the articles, I listened to the presentations this afternoon, I still don't understand it. I think that the details left to work out are huge. And that's probably because I haven't done it yet. There are fuzzier things—and I don't think that these are as amenable, but if somebody tells me that they are, I'll rejoice. [*laughter*]

I'm not saying take my word for it; I'm just saying that there are things where I have more doubts.

Now, my last slide is, I think, the most key slide of all: my questions. What form of specification? Because I think the thing that the computer scientists ought to know is that our specification is often a Word document. And somebody has described, in a Word document, everything. Valid values, flow, question text. And that does not seem amenable to model-based testing in the sense that you're trying to generate test scripts from your specifications. So, some sort of database specification seems to be required, but what is the minimal information content? And one of the presentations just before this one got at, you know, putting some information in your database about what you should expect. But, still, I don't have clearly formed in my mind how that database is going to look, for both specification and some of the results. And then there are a lot of nicely-named algorithms: where do we find out about them? More importantly, is there yet a textbook about this? I mean, what is the maturity of model-based programming? I know that there's a textbook on extreme programming, or at least a book. But is there one on model-based testing, because the articles I read—to be quite honest—extol the benefits of it and give some little examples but don't really get into the nitty-gritty about how to do it. So, that's it.

Yes, Harry?

ROBINSON: I guess I can answer, I think, the last one of those, about textbooks. About the actual automation part of it, there aren't yet. But there is a book from 1995 or so called *Black Box Testing*. In fact, what they do in that, their running example is the 1994 1040 form.

SMITH: Just a comment . . . I liked the slide you had on all the other things that come into play and are really relevant for some applications. In computer-assisted instruction, a huge amount of human time goes into the media and GUI. . . . I wonder whether these surveys have a major GUI portion.

PIERZCHALA: They can. We do these, this kind of interviewing where we turn the laptop around to the respondents and play, you know, radio ads—"have you heard this ad against drug abuse?" Or we play a clip from television against drug abuse. There's more and more of this kind of GUI aspect to it. But most of it is still an interviewer reading a question.

McCABE: I'd like to extend the notion, a little bit, about model testing. I've seen projects where the specification might be a Word document, a narrative document. But the exercise of coming up with scenarios that would provide a test can work off a Word document as well. And a lot of times it's thought that you're testing a product that's built, but what's missed is that it's often useful to test the requirement. For example, the requirement might be that the system be interactive. Well, what's the test for that? You put the input in, and it comes up three days later, does that show that it's interactive? But to specify the interval of time, for example . . . So what you find is that when you take any kind of document—narrative, state shell, whatever it is—and start fleshing out the tests, it often shows that the model itself is incorrect in the system, way ahead of the product. And the process of refining the test, and getting consensus that this is a test, it's complete and robust and accurate, often in a project is very worthwhile, whereby you debug the specification in whatever form it's in. And then you have consensus before you move any software or survey to apply those acceptance tests.

PIERZCHALA: Certainly, one of the things I advocate is to get test scenarios set very early.

McCABE: My point is that it's not so much technology-based as it is methodology-based.

PIERZCHALA: I understand that. I was reading in the literature where it says that these things are generated; what you're saying is that you can also apply the methodology even if things aren't fully in a "generable" form.

McCABE: And you can do better.

PARTICIPANT: Mark, I also liked the fact that you reviewed for everybody that there's a lot involved in the testing that isn't going to be addressed through model-based testing. You still have to do the screen standards and the cosmetic, the navigation and the usability, all of which is very time-consuming and expensive. So I have a question for any of the presenters who talked today about model-based testing, and that is: if you have a client who gives you a project that is both time- and budget-constrained, will the model-based testing be cost-neutral? That is, do you save enough money from it, in perhaps avoiding some of the more iterative phases of the traditional sorts to testing, so that it would not cost the client any more?

ROBINSON: I can take a swing at that. We have a lot of teams that are light on budget, and their refrain is, "we can't afford to do that." And our response has been, "you can't afford *not* to." Because what they're doing is, it looks like they're saving money up front but they'll actually be paying for it later on.

PIERZCHALA: Let me just say, one of my jobs this afternoon, as given to me, was to sort of pull everybody back to earth. And I think I did that a little bit. I would say that I like the idea of model-based testing, and I can even see myself *applying* it once I've learned more about it, how actually to do it, because I think I'll eliminate the surprise factor—these combinations of data that nobody's going to put into a scenario but that are going to blow up a calculation, for example. If I can get rid of that kind of stuff, I'd be very happy. Yes?

SMITH: Just to respond to the question. I think that given a very appropriate sort of model, sort of Markovian model-based testing we were doing at Computer Curriculum, I think everyone would agree that it really saved a lot of time and money. There would have been no question about that, though we never tried any scenarios to see. Then we were getting killed by the other kinds of testing in this slide show. But we saved enough money with the model-based testing to test the GUI.

McCABE: I have two comments. One is, you think about the process and think of getting a model early and a test early, and the irony is that a model, or a Word document, is often as wrong as the software. It's inconsistent, it's incomplete, it's not regular … So when you think of the test, an acceptance test to apply to the spec itself, usually it'd be worth it. Now the cost of those errors multiplies if you don't catch them earlier, so you're catching very expensive errors early. That's the good news. The bad news is that this often doesn't work because of time frame. See, what gets in the way is agencies having an RFP before they have any requirement document. And the problem is that the contract is bid on, and the bid is for a fixed price and etc., and *no* one would go back to the sponsor and say, "these requirements are all wrong." Because they want to get paid.

DOYLE: And they want to win the contract.

McCABE: A lot of lifecycles go wrong because there's a conflict, if not a conflict of interest then at least a conflict of problem. And it gets in the way of everything. But, otherwise, it saves as much money by avoiding doing the wrong thing.

PIERZCHALA: Yes, Harry?

ROBINSON: Just back to Tom's comments on Word documents and such … What we've been doing is that the Word documents that are written up by our system engineers, those have been converted to models on the test side. And typically you run into what we call "specification rot," because you kind of use the use cases and then nobody uses them again and they drop out. But we use our specs to drive the tests so that—rather than being a requirement that somebody keep the specs up to date—it actually serves our purposes to have them up to date.

PIERZCHALA: Yes, Mike?

COHEN: There is a textbook coming out on April 26 by a student of Jesse Poore's, James Whittaker.

PIERZCHALA: Oh, good … did everybody hear that, on April 26, rush out to your local bookstore … Who's the publisher?

ROBINSON: Addison-Wesley, I think.

PIERZCHALA: I'm glad to hear that because textbooks can gloss over some issues but they also tend to bring issues together and synthesize them better than just going one article after another.

ROBINSON: Just because I can't let something go, that model-based can't test something … [*laughter*] Just going on a thought here. Getting back to the problem of getting the grammar right, the fills, would it be feasible—if you're going to generate something that will take you through all of the questions—that you could then automatically put all the questions through a grammar checker?

PIERZCHALA: I think that there are ways to do that because, after all, those fills are just a variable or a field. And I think semantically it's harder to tackle but I'm not willing to say it can't be done, either. In fact, I would really love it if it could be done.

DOYLE: That would take a lot of work off . . . we would *love* that.

PARTICIPANT: I think that you could use part of the model-based testing and combine it with the human factors, present these cases for review. Our difficulty is hammering the keyboard and getting it to a place where we can look at it again, that you could use this technique early on to identify cases that you want to take a look at and look at the human factors and the fuzzy concepts you identified.

GROVES: Let me make sure I've got it right . . . Does model-based testing integrate its own notions of complexity as priors on the model? So if you had a region of activity which was very complex . . . that could be subjected to more testing?

ROBINSON: It's probably been done; we've never actually done that, but what we've done is that—as you make up the model—you suddenly begin to realize that all of the arrows go into one spot. Or there are way too many arrows coming out of one. So you look at it and say, "that's untestable." It would be wonderful to have some sort of complexity you could just run on the rest . . .

McCABE: If I could add to that . . . I did a company's research that built some stuff. At that time, there was a thing called "data flow diagrams" (DFDs), and I developed some mathematics that could develop a test based on those diagrams. And the beauty of that is the requirements specification and the test specification were the same. So what would happen is that the agencies would develop DFDs as a specification, and we would derive a test based on that. Now the interesting thing is that if you [take a walk through the program and compare with] the DFDs, about 30 percent of them are off because they've never been tested. Because the DFD can lead you to places, and it never went there. And the way you'd find that out is that you derive the test, and you find things—for example—in the DFD like sections you can't get into. . . . So we did that for a couple of years, published some things. Now there's another set of work related to that, and that's called "work flow testing," developed by Musa at Bell Labs and used there and other places. It's very much like what Harry's talking about in that it adds what are called "characteristic nodes." So you can take the clock, for example, and it might have a lot of usage scenarios, but it might be that scenarios one and two are used 90 percent of the time. So out of that [you can get a sort of stress test.]

But . . . the fundamental idea is to test up front, whatever the medium is, and to build the requirements in.

DOYLE: The diagram you gave; is that some kind of software that could be run on our instruments, to see how complex it is? Highly complex? Somewhat complex?

PIERZCHALA: I didn't see either CASES or Blaise as one of the 17 supported languages ...

McCABE: We didn't commercialize that. Now the interesting thing is, you know why we didn't? Because people don't do it ... They've got all the stuff, for the programming languages; that you could do, because the programmers will test. But we couldn't enough [other] people to test early. I mean, it sounds insane, but that's the reality. So I just wanted to kind of get you guys off the defensive, in surveys, because software has all these problems, too. We're kind of birds of a feather. The problem is with these communities is that you talk to people, and they don't want to fund testing. Microsoft does very, very well—they're probably atypical—but in a lot of organizations testing is a poor cousin. You don't get the money, the funds, and so forth. So we have the same problems you do. We haven't commercialized that work because there wasn't a big market, enough market ... Now we did use it in some security, very high-risk environments ... And just like software it's never right the first time; the requirements are never right until four or five iterations, like testing.

CORK: Thanks to everyone for coming. I think that was a great first day.

GROVES: When does the evening session start? [*laughter*]

CORK: We look forward to seeing most, if not all, of you again tomorrow morning.

WEB-BASED DATA COLLECTION

Roger Tourangeau

CORK: Welcome back for Day 2 of our workshop. In Day 1, we concentrated on documentation and testing, after getting an overview of the general problems. For Day 2, we wanted to do something a little bit different; in this first block of talks, we look at some other emerging technologies coming around the corner, areas in this nexus between survey research and computer science where further interaction could be useful.

To start off, to talk about Web-based surveys in general, we have Roger Tourangeau, who is director of the Joint Program in Survey Methodology.

TOURANGEAU: Good morning ... I'm going to give a kind of fast review, in 25 minutes or so, of issues in Web-based data collection. I'll look mainly at measurement issues, and a lot of what I'm going to tell you is information I stole from Mick Couper. So, if you disagree with most of it, hey, talk to Mick after the talk ... The Web has really caught on, and it looks like an incredibly low cost means of survey data collection. So, the average face-to-face interview now costs, what, maybe $750 or $1,000, on average, to do a face-to-face interview in an area probability sample. It's literally true—you can get, for $2 or $3 dollars a case, an interview on the Web. So lots of firms, particularly in the market research community, have rushed to adopt this new technology. Yet, I think we all think—those of us who have looked at it carefully— for government applications and high-end academic applications there's a lot of research that's going to be needed before federal agencies accept the Web.

And why is that? Well, there are several key issues involving sampling and coverage, non-response, and then there's this host of measurement issues.

What are some of the sampling issues? For sampling purposes, there are two key obstacles for Web surveys. For most survey populations likely to be of interest to the federal government or academic survey researchers, there's no good frame. There's no frame of Internet users. And even if there were a good frame of Internet issues, it's not clear ... there are significant coverage problems; only a fraction of the population—a *growing* fraction—has Internet access. So for most surveys there would be huge coverage problems.

Because of these problems—the absence of a frame and the presence of these significant coverage problems—there have been four methods of sampling that have been adopted, mostly by the market research community. There are what might be called general invitation samples, volunteer panels, probability samples, and then—as a special case of probability samples—intercept samples, and I'm going to talk briefly about those four.

The general invitation samples are simply Web sites where you can go and fill out a survey, if that's your cup of tea. I love this; there's, for example, a Web site [called] MrPoll.com, and you can go there and register your opinion on various issues, at no charge to you. [*laughter*] These are the moral equivalent of those 900 polls on TV, except that they're even cheaper to do, right? When you do a 900 poll, you actually pay a few bucks to participate. Here, it's free. It's a completely self-selected sample, so it's a sample of convenience; it has unknown sampling bias and it's obviously restricted to the Internet population. So, not a really good sample.

There are also volunteer panels. The largest of the volunteer panels— at least the largest of the volunteer panels that I'm aware of—is the Harris Interactive panel. I think they're up to several hundred thousand members of the panel. And you can get a probability sub-sample of this large volunteer sample. So you can get a probabilistically cleaned-up version of a really dreadful sample. Again, the coverage is ... so it's not really a probability sample, and it's coverage is obviously limited to Internet users. [George] Trehanian and his various colleagues at Harris Interactive have attempted to improve the quality of the estimates from the Harris Interactive panel by using propensity weighting, where they're predicting the likelihood that you'd be in this panel as opposed to their telephone sample. They've had parallel telephone samples. And so they calibrate the results from their Web panel to their telephone surveys. And, in the presidential polls, they were dead on; they did pretty well. Again, though, I'm not sure how many of us are ready to trust the miracle of weighting as opposed to probability sampling.

There are some probability samples used for Web surveys. It is possible to get a list sample—for example, all the e-mail addresses of students at a single university. My colleague, Mick Couper, has done several studies like this; he's surveyed students—where is Mick? All right, oh, there you are, you're hiding. He's done several surveys of students at the University of Michigan, for example. A number of studies have offered the Web as one means of responding among other means, so the sample is not restricted to Web users. For example, I worked on a study—the National Survey of Post-Secondary Faculty, in 1998—where respondents were mailed a questionnaire and invited to respond via the Web if they wanted to; they could also respond by phone. So the Web is just one option among others in this probability sample. These samples are typically ... Some samples are identified through other means—for example, random digit dialing—and then asked to respond via the Web. In principle, these samples present no new sampling problem: you have some frame, you do probability sampling, you do your usual thing. Then you ask people to respond by the Web. And the problem there is that you can have much higher non-response as a result. For example, in Inter-Survey's panel, people are contacted by RDD; there's a certain dropout at that stage.[47] They are asked if WebTV can be installed in their home; there are dropouts at that stage. After a while, people drop out of a panel; on any particular survey, you always have non-respondents. When you multiply ... even if you had an 80 percent response rate at all those stages, you're down to 33 percent when you multiply .80 times 4. And in fact they don't have 80 percent at all stages; their response rates are typically in the 20s.

---

[47] InterSurvey was a Palo Alto, California, company established to conduct surveys using the Internet. It is now Knowledge Networks, Inc., and more information on the firm is available at www.knowledgenetworks.com.

And then the final approach is that you can kind of intercept people at particular Web sites. So you can get a probability sample of visits to a particular Web site through this means. There are various ways you can do this; in some Web sites a banner ad pops up, and then you can go to another Web site to do a survey. There, obviously, you don't get a very high response rate by this method. There are pop-up surveys; sometimes, the questions just come up. And then you have to get out of it if you don't want to do it. And then there is hijacking, where the survey takes over the browser, holds your children hostage, [*laughter*] and you have to complete the questions before you can move on. Often, it would be difficult to calculate a response rate in these intercept surveys; you'd have to keep tabs on the number of visitors. And the response rate is likely to be low. Another drawback to this approach is that you're getting a sample of visits, not visitors, right? So the questionnaire comes up every *n*th time and if it's the same person—like Mick [Couper] is a habitual Web site abuser and might visit the same site 20 times a day or something—you'd get Mick 20 times if you weren't careful. Although he has plenty of different opinions, so there would still be good variance properties ... [*laughter*]

If you want to use the Web to survey, to make projections to the general population, there are likely to be difficulties. Not only is the overall coverage less than perfect, there's differential coverage by lots of different characteristics. For example, in education, if you have a college degree or more, 82 percent of those people—according to the Pew Internet Project—have Internet access. Whereas it falls to 37 percent for people who only have a high school diploma. Likewise, it varies by age; the elderly population, 65+, are less likely to have access. It varies by income. It varies by race, as well. So some groups approach reasonable coverage—high income groups, college graduates. But lots of groups have pretty poor coverage.

You probably can't read this very well, but I'll take you through it quickly. So one set of issues is: it's hard to sample in the absence of a frame, and, if you want a sample of the general population, there's severe coverage problems. A second set of problems with Web surveys is that the response rates tend to be low, and this slide—which I stole from Mick—shows the response rates in various mode comparison studies. [*See Table II-2.*] And I'll just take you through a couple of studies. Radler has done a number of experiments on University of Wisconsin students, and he's compared mail survey with Web, and in both cases you get a much higher response rate with mail. Guterbock finds the same result. There was another survey done at the University of Georgia with similar results; the mail exceeds the Web. It isn't *always* the case that the response rates are dreadful. Bates did a study of U.S. Census

Table II-2 Response Rates from Explicit Mode Comparison Studies (List-Based Samples)

Source	Target Population, Topic and Design	Response Rate by Mode
Kwak and Radler (2000)	University of Wisconsin students; 1999 survey on campus computing; $n = 1,000$ per mode	Web: 27.4% Mail: 41.9%
Radler (2000)	University of Wisconsin students; 2000 survey on campus computing; $n = 1,000$ per mode	Web: 28.0% Mail: 52.6%
Guterbock et al. (2000)	All computer users at University of Virginia; survey on university computing; $n = 549$ Web, 362 mail	Web: 36.8% Mail: 47.6%
Bason (2000)	University of Georgia students; survey on drug and alcohol use; $n = 200$ per mode	Web: 15.5% Mail: 27.7% IVR: 17.4%[a] Phone: 23.9%
Crawford et al. (2002)	University of Michigan students; survey on alcohol and drug use; $n = 3,500$ per mode	Web: 61.6% Mail: 40.6%
Jones and Pitt (1999)	Staff at 10 English universities; $n = 200$ Web (email invitation), 200 email only, 100 mail	Web: 19% Email: 24% Mail: 72%
Weible and Wallace (1998)	Management Information Systems (MIS) faculty in U.S., from directory; $n = 200$ per group	Web: 26% Email: 24% Mail: 35% Fax: 25%
Bates (2001)	U.S. Census Bureau employees; organizational climate survey; $n = 1,644$ Web, 1,645 paper; invitations by mail	Web: 66.6% Mail: 62.8%
Raziano et al. (2001)	Geriatric physicians; $n = 57$ per mode; mode reversed for follow-up; initial response rates reported	Web: 39% Mail: 63%
Klassen and Jacobs (2001)	Establishment survey from directory of Canadian firms; firms contacted by telephone to get names and contact information; thereafter randomly assigned to mode; $n = 110$ Web, 163 mail, 161 fax, 160 DBM[b]	Web: 14% Mail: 23% Fax: 20% DBM: 21%
Vehovar et al. (2001)[c]	Survey on e-commerce among Slovenian companies, in five experimental groups: *A*: solicited by phone, respond by phone ($n = 900$); *B*: solicited by mail, respond by mail ($n = 300$); *C*: solicited by mail, respond by fax ($n = 100$); *D*: solicited by mail, respond by Web ($n = 300$); *E*: solicited by email, respond by Web ($n = 200$)	*A*: 63% *B*: 54% *C*: 43% *D*: 26% *E*: 7%

[a] Interactive voice response (IVR) surveys are automated telephone surveys; respondents answer survey questions through verbal response or by pushing keys on the telephone keypad.

[b] Disk-by-mail (DBM) surveys involve sending respondents computerized questionnaires on a diskette, which are completed by the respondent and returned by mail.

[c] The figures reported for this study are completion rates, not response rates. The completion rate is the proportion of respondents with respect to the entire sample while the response rate is with respect to the number of eligible participants. In this case, the completion rate denominator for group *D* would include those firms for whom Web access was not available; likewise, the completion rate denominator for group *E* would include firms with no listed email address. These quantities would be excluded in computing response rates.

SOURCE: Workshop presentation by Roger Tourangeau.

Bureau employees and actually got a higher response rate, and not a bad response rate—66.6 percent—in that kind of a setting. But generally what you are seeing is low response rates. And, in part, that may reflect an absence of an empirical basis for knowing how to get a high response rate on the Web. This is a new mode of data collection; we don't know what the right schedule of follow-ups is, how to give advance notice, and so on; all that technology which is well established for mail surveys and telephone surveys isn't there yet for the Web. So the techniques for increasing response rates on the Web are still in their infancy, and you can see it in the poor response rates.

There are actually some perverse results that suggest if you add the Web as one option among several in a mixed-mode survey you can wind up with a slightly lower response rate. So people say, "OK, I can do it on the Web," and they never get around to doing it on the Web or they break off on the Web, or something else happens, and you end up with a worse response rate with the Web as an option than if you hadn't. That's a result, by the way, in the ACS—the American Community Survey.

Another problem, non-response related problem, with Web surveys is—aside from ordinary unit non-response, "hey, I don't want to do it"—it's awfully easy to break off in a Web survey. So you do a few questions, and you hit despair after a while . . . and so you stop doing it. And it's a lot easier to break off in a Web survey because there's no interviewer there to say, "Oh, *wait*, don't break off, don't be rude to me . . . ." And a series of studies by [Mick] Couper and colleagues suggest that sometimes progress indicators are helpful; these are the little bars that say that you're partway through—"go, go, you're partway through." But if the news is bad then people become even more discouraged.

Despite the nonresponse problems I think that you're going to see the Web more and more as an option in mixed-mode surveys. And, for example, the ACS offers a Web response option, the American Community Survey. And people are thinking about this as an option in Census 2010; I don't know if the Census 2000 allowed a Web option . . .

PARTICIPANT: The American Community Survey doesn't have a Web option . . . We did a Web test for the American Community Survey.

TOURANGEAU: I stand corrected.

DOYLE: Census 2000 did have a Web option, but it wasn't advertised . . .

TOURANGEAU: A carefully guarded secret; I like that.

DOYLE: If you looked at the letter, you could see that there was an option . . .

TOURANGEAU: And, of course, no one in America—this being America—actually got to the end of the letter [*laughter*] where this spec was revealed.

OK, I want to spend the remaining 15 minutes or so of my time talking about measurement issues. And, relative to familiar methods of data collection, the Web offers an enhanced capacity for visual material, for audio material, and for interactivity between the interface and the respondent. All these possibilities—which seem so exciting—also raise some concerns. And so we've started—Mick and me, and Reg Baker, and Fred Conrad, and some others—have started doing some work looking at these problems, with support from the National Science Foundation.

There are three issues that we think are especially important or, at least, when I was doing this talk, I thought, "these are the three things I'll talk about." One is the overall set-up of the questionnaire, its format. Tradition has already evolved that says that Web surveys are just like mailed paper questionnaires, and we should transfer that technology to the extent possible. So the rules for creating an ordinary [self-administered questionnaire] should be followed in constructing a Web questionnaire. And that's what I call the "static format." And the other tradition that has emerged is that Web surveys are like computer-assisted surveys, so they ought to follow the conventions that CAPI surveys and CATI surveys follow. And I'll talk about that a little more. Then there are some issues about what sort of interface you should have and whether you should add humanizing features to the interface. Finally, there's concerns about the impact of visual images, photographs and the like.

Here's an example of what I call a static questionnaire. [*See Figure II-24.*] This is our attempt—Mick and I and Reg Baker are involved in this project—to take a questionnaire used as a part of NAMCES, the National Ambulatory [Medical] Care [Evaluation] Survey—and make it part of a Web survey. Clearly, what we've done is follow the conventions that the paper questionnaire follows. For example, instead of having a separate screen where you would get a follow-up item, this sort of directs the respondent over to this follow-up item. OK, and likewise, 3b follows that same convention. So it looks like a paper questionnaire; it's just been transferred to the Web.

Alternatively, you can imagine presenting item-by-item the questions, and then skipping automatically; if the person answered "yes" to 3a then they would never see this. This is the way we would do it in a CAPI survey.

There's not a lot of empirical literature; there have been three or four empirical tests that have compared the interactive format with the static format, and for the most part they don't demonstrate a lot of differences between the two. In theory, there are some advantages and disadvantages to each approach. It's easy to program the static version, it minimizes download time, it accommodates low-level browsers, the respondent can view the entire questionnaire. On the other hand, skips

3. CONTINUITY OF CARE

a. Are you the patient's primary care physician?

☐ Yes

☐ No   } **Was patient referred for this visit?**

☐ Unknown   ☐ Yes

          ☐ No

          ☐ Unknown

b. Have you or anyone in your practice seen this patient before?

☐ Yes, established   } **How many past visits in the last 12 months?**
patient
          ☐ None

          ☐ 1-2

☐ No, new patient   ☐ 3-5

          ☐ 6+

          ☐ Unknown

Figure II-24   Example of static design for a Web questionnaire.

and edits can be difficult to handle, you can't control the order of completion, you get more missing data, if the system crashes the whole thing is lost, randomization is difficult, and so on. I'm not going to walk you through the rest of those.

So that's one issue—how do you set them up?—and there's not much empirical literature; there seem to be some a priori pros and cons to both approaches. A second issue is what kind of interface do you build into a Web questionnaire. Already, there's been—again, given that many of the samples consist of volunteers, there's been an effort to sort of jazz up questionnaires by adding humanizing touches, making the questionnaire more entertaining. And here's an example of the Lightspeed MiniPoll, where you have this attractive young woman inviting you to take part— "it's *fun* to voice your opinion, and compare it to others!" There she is again—"Thanks! That was easy!"—and then they're inviting you to be in a LightSpeed panel. [*See Figure II-25.*] And so the idea is that a lot of Web surveys add various humanizing touches to make the survey more fun.

There's been a small industry in research—mainly generated by two groups of researchers at Stanford and Carnegie Mellon—who have looked

Figure II-25   Humanizing touches in a Web questionnaire interface.

at the consequences of adding humanizing touches to computer interfaces. And their results suggest that humanizing an interface can create problems. People begin to react to these interfaces as though they were interacting with a person. So, for example, in the survey context, you may get social desirability bias effects just like you would in interacting with a human interviewer. This is especially true when social interface features like voice or visual cues are added. And this stands in stark contrast with the survey literature that says when you add voice, for example, to Audio-CASI, that actually improves things. So voice presumably is one of these humanizing cues, but it nonetheless seems to generate improved reporting. Their work is based on laboratory experiments with student subjects.

And here's an example; this is a study by Nass, Moon and Green (1997), who are part of the Stanford group of researchers. They had students interact with a tutor computer—say that three times quickly—

and then evaluate it on a second computer. So they interacted with one computer that was trying to teach them something and then evaluated it on a second computer. This is their ratings of the tutor based on features of the evaluator computer. And you see that the voice on the evaluator computer seems to have a significant impact on the assessment of the tutor computer. And they conclude that there's a tendency to gender stereotype, and it can be triggered by minimal cues such as the voice on a computer.

PARTICIPANT: Can I just ask, was the voice on the computer a male or a female?

TOURANGEAU: On the evaluator computer some got a male voice and others got a female voice, and that's the difference.

PARTICIPANT: Which of those is the tutor computer and which is the evaluator computer, the one on which the voices varied?

TOURANGEAU: I think this was the voice of the evaluator computer, so this is mis-labeled. And these are the ratings of how friendly was the tutor computer and how competent was the tutor computer based on the evaluator computer voice.

So, Charles Turner—who's looked at Audio-CASI—has argued that, even in sex surveys, the gender of the recorded voice is unimportant. Whereas Nass, Moon, and Green argue that the voice selection is highly consequential; a male voice brings with it a large set of expectations and responses, blah blah blah.

Mick and I have done three Web experiments that compared a neutral interface with interfaces that featured human faces and various other interactive features. And we were looking at the various possible survey response effects that might emerge when you use this humanized interface, including people changing their answers to gender-related questions to appear more feminist if they had a female picture. Or differences in social desirability bias.

So, there's an example of what we did as a humanized interface … [*See Figure II-26. Laughter; face in example is speaker's own.*] This was one of the major sexes; I can't remember. And I don't know if you can read this, but this is me giving feedback to a fictitious respondent: "According to your responses, you exercise more than once a day. The fat you consume is much less than average, and you currently weigh pretty close to your ideal weight. Thank you for providing this information … you liar, Mick." Here's the female investigator, Darby Miller-Steiger—I'm one of the investigators on the project. And then [there] was the neutral interface.

We did a series of three studies; those three pictures, I think, were from the first study, male, female, and the logo were contrasted. There was a high level of interaction. Across—I'll just summarize—across the

Figure II-26   Web instrument with human face added to personalize the instrument.

studies, we don't find a lot of differences in the degree of socially desirable responses. We gave people a couple of standard batteries of items that are supposed to measure social desirability. The Marlowe-Crowne items, and something called the Balanced Inventory of Desirable Reporting, BIDR.[48] We didn't find any differences across versions of the interface, nor did we ... we also gave a battery of about a dozen survey items that asked about things like drug use and drinking and diet and exercise, and so on; voting, church attendance, sort of standard items that are known to be susceptible to social desirability effects in surveys. And we don't find a lot of differences. Where we do find differences is in gender attitude responses. And the higher numbers here indicate more pro-feminist responses. In each case, the attractive fuchsia bars represent the female interface and the blue bars represent—guess which major sex? And then the grey bars are the neutral interface. And you always see this trend towards more pro-feminist responses with the female interface. So there is some evidence suggesting that adding humanizing cues to an interface can change things a little bit; we find little support for the social presence hypothesis, this idea that people react to an interface as they

---

[48]The metrics referred to here are the Marlowe-Crowne Social Desirability Scale (Crowne and Marlowe, 1960) and the Balanced Inventory of Desirable Reporting (Paulhus, 1984; also known as the Paulhus Deception Scales).

do to a person in response about sensitive behaviors. But we do find some results for gender attitudes, perhaps due to a priming effect.

I'm just about done here. The final issue I wanted to talk briefly about involves the effect of images on the answers to survey questions. One of the appeals of the Web is that it's very easy to present visual images—video clips, photographs, drawings, the whole bit. On the one hand, you can see how these are appealing, because they could really help clarify the questions. But we were concerned that they might carry misleading implications as well; in particular, images are necessarily concrete. So we thought that images could affect the construal of the target category. And let me give you an example; this was our study design. This was a study that Mick and I did with a researcher at InterSurvey. We asked people how many overnight trips they made over the past year, and some people—with that question—had a picture of a businessman at an airport. Some people got a picture of a family loading up a station wagon, getting ready to go on a trip. And some people got both pictures, and some people got neither picture. The idea was ... what we called the Activity A pictures were highly salient—but not necessarily common— examples of the category. So people attend more Little League baseball games than professional sporting events, but they're more likely to think of a professional sporting event when you ask, "Did you go to a sporting event last year?" You're not likely to think of the neighborhood Little League. At least that was our hypothesis. And here's an example of the stimulus: "Overall, how many sporting events have you attended in the last year; that is, since April 2000?" This is the Little League picture and it's a little bit better on the Web but you get the idea.

These actually were our two best examples—the sporting events and the shopping—where the data conformed best to our hypothesis. So with the salient instance—in this case, the professional sporting event— people reported 2.7. With no pictures, 3.1. With the Little League picture, 3.7. And with both pictures, 3.4. Likewise, here, we asked people about shopping. The salient instance was ... gee, I can't even remember. A department store. And the non-salient instance was a grocery store. You show people a picture of the department store and they report 7.7 instances of shopping; 8.7 with no picture; 9.0 with a picture of the grocery store; and with both pictures it goes up to 9.8. There seems to be some kind of impact of the image on how the category is construed, and it affects the overall frequency.

So let me wrap up. There are difficult sampling issues in stand-alone Web surveys—Web surveys that don't offer other response options— particulary when they intend to represent the general population. The situation, the sampling situation, is better for certain restricted populations—for example, students at a single university. Mick and I have been investigating the possibility of doing surveys of physicians, and it looks like 95 percent of physicians have Web access. The situation is also better for mixed-mode surveys. Response rates are generally low, but they can approach 70 percent. There are many, many tough measurement issues. Some of these design issues reflect that there are two separate and contradictory design traditions: one says that Web surveys are like usual computer-assisted surveys and the other says that, no, they're like typical mail surveys, like paper questionnaires. In addition, the visual character of Web surveys raises some interesting issues.

That's it; thank you very much.

All right, you're allowed to ask questions now. Pat?

DOYLE: I'm curious that you would present the response rates as sort of being OK, because the experience in what I've seen with household interviews is that they really are *very* low. And, in fact, in the numbers you put up, the only number that was above 60 was the Bates study, which was not a household survey; it was a survey of people in businesses. It does seem that the Internet surveys are much more successful on the business side than the household side. I'm real surprised to kind of hear you tout those response rates as encouraging.

TOURANGEAU: I think that, in time, I can imagine that these rates will approach those of mail surveys. So that's maybe the right, guarded conclusion. You can imagine Web surveys using a sequence of reminders, similar to what's done in mail surveys, and achieving similar results. Now, for the general population, absolutely not. I mean, right now you don't really have access to the general population unless you do special things like install equipment in their house. So ... I know, I think that for the same kinds of applications where a mail survey might be reasonable, you might be able to get similar response rates using a Web survey.

DOYLE: But you're not—I guess the numbers don't show that yet ...

TOURANGEAU: In the one case, it did ...

DOYLE: ... unless you're really optimistic. It was only in the case of the business study where the rates were comparable, the Bates study, where literally everyone was at work.

TOURANGEAU: Right. But isn't it the case, though, that mostly mail surveys are done when there's a list frame, and so on? So I think that's the feature, and not the business thing. Miron?

STRAF: Your visual images are really just setting the context for what you mean by shopping or sporting events, and you don't expect the same results if you add language? Such as, "a Little League game"?

TOURANGEAU: You know, we've gone back and forth ... that's a good question. And I definitely agree with you, your first point, which is that I see images as context, as survey context. Part of our motivation for this study is: because images are so easy to do on the Web, you see a lot of it. And the examples may not be particularly well-selected. So that's an issue. There is some literature suggesting that people are more likely to process images than they would verbal material. But in principle what you've said is right: you've drawn some exemplars from the category and it helps shape the perception of the boundaries. And in the case where you do the highly salient instance it seems pretty clear that they're drawing the boundaries more narrowly than they would if you take a non-salient instance.

You had a question, I think?

PARTICIPANT: I'm just curious about the difference in response rates between the mail and the Web. Most people I know are somewhat more interested in using the Web in those instances. How do you ... [*trails off to inaudible*]

TOURANGEAU: Well, in every study but that one, mail exceeded Web in terms of response rates; that was the point Pat was making. In every single comparison, mail exceeded Web, significantly so. And so, I think, though, that mostly reflects our inexperience with Web surveys and the difficulty. I mean, maybe one big difficulty is that it's easy to slip a two-dollar bill into a mail survey but it's more difficult to do that ... I think that none of these studies—Mick, correct me if I'm wrong—but I don't think that any of those studies used incentives, did they?

COUPER: There is one recent exception, a study that Reg and I are involved in. Again, a survey of Michigan students where we got a 63 percent response rates by the Web versus 41 percent on the mail. And what we did is deliberately send out—this is to students at Michigan—a Borders gift certificate; Borders is headquartered in Ann Arbor. And when they completed the Web survey they got an ID and could immediately claim a $10 gift certificate.

TOURANGEAU: Was that in both groups, or just the Web?

COUPER: It was in both groups, but was more of a factor in the Web group.

TOURANGEAU: So, again, my conjecture—and it is just a conjecture—is that with greater experience we'll get to a position where Web is sort of the equivalent of mail. Yes?

PIAZZA: I was talking recently with the person at Berkeley, with the person who does surveys of the student population there. And he said that they used to do these surveys by mail but that the response rate had dropped way below 50 percent. And now he's doing all of his surveys by Web, with some e-mail follow-up, and he routinely gets about 50 percent—a much better response rate by Web than by mail, right now.

KALSBEEK: So does that possibly mean that the convergence that you anticipate—and I tend to agree with you—might be more a matter of the technology generation getting older and beginning to move into the population. Is age a factor in the things Mick is talking about; do you expect these effects more because use of these tools is more of a way of life for younger people?

TOURANGEAU: I like that speculation, too ... I would anticipate convergence, or even a cross-over, happening in the younger age groups. Yes?

PARTICIPANT: What is the length of these Web surveys before you start to see a lot of break-off?

TOURANGEAU: I don't think that there's a magic formula here. I think a lot of it has to do with what you anticipate. For instance, we did a study—this is actually using interactive voice response, telephone Audio-CASI. Well, actually, we used the long-form questions from the census, the decennial. And people broke off whenever they got to a new person; you know, a light went off, "[Uh-oh], 35 dull questions about to come." And then they hung up. And I think you'll see similar phenomena when it comes to the Web; when people can anticipate a lot of tedious questions, you'll have break-offs. On the other hand, I've also done some surveys where you had no progress indicator, and it was just one item after another, using the interactive format, and you're just totally clueless about how far along you are, and I—at some point, that's not a successful strategy either, and there, it just might be a function of time rather than anticipation. Miron?

STRAF: If I could ask another question, on the collection of *sensitive* information ... some of what you're showing seems to show that there aren't major desirability effects and the like. But are there alternatives, and here I'm thinking of things like randomized response, using the Web?

TOURANGEAU: I'm not sure. All of our results tend to come from comparisons of different Web interfaces. It's conceivable that people didn't, weren't sensitive to the different interfaces because in every condition they're underreporting sensitive information. You know what I mean? In fact, there's one study done by Moon, one of the Nas collaborators, who asked the same questions but portrayed the computer that the respondents were using as either a standalone computer, not hooked up to any kind of network, or part of a network and the questionnaire resided on a computer across town. Or on a network and the questionnaire resided across the country. And you got more socially desirable responding when it was viewed as being across country, hooked up to a computer across country. And it might be that people, when they're using the Internet, are very sensitive to the fact that people could be eavesdropping, that their information isn't very confidential. So it could be that what we're getting is that everyone is lying, under every condition. And it's just not clear if it's uniformly bad or uniformly good reporting in our studies. But it's uniform; it wasn't affected by the interface.

OK, thank you very much.

INTERFACE OF SURVEY METHODS WITH GEOGRAPHIC INFORMATION SYSTEMS

Sarah Nusser

CORK: The next emerging technology that we're going to discuss is the interface and uses of global positioning satellites (GPS) and geographic information systems (GIS) and their potential uses in survey research. To talk about those, we have Sarah Nusser from the Department of Statistics at Iowa State.

NUSSER: I'm going to talk today about digital geospatial data and survey data collection. We've actually been using spatial data in the survey process for a long time; it's just been in primarily paper forms. And yesterday we talked a lot about the process of developing software; I'm talking more about the planning, navigation, and data collection side of the survey cycle. And I just wanted to give you a little context for why I'm working on this. I work on the National Resource Survey, sponsored by USDA, that uses a lot of geospatial information that has historically been in analog or paper form. And they do both photo interpretation and field studies, and over the last five years, we've been using—and, thank you, Marty [Meyer] and Jay [Levinsohn]—Apple Newtons to develop the data collection software.[49] And there's a client-server setup to send out information to these people, have them collect it and send it back in.

But we had not been working with spatial information. Our first foray was back in 1999, when we sent people out into the field with GPS units. And we couldn't get precise positioning signals back then, so we had these great big military plugger machines. And you can see the interface on the GPS machine is 4 lines with about 20-some characters. And we found people made gross mistakes in how they set up the equipment and then how they used it to capture spatial information. So what we did was connect the two together and develop an interface that allows them to see what's going on in the guts of this machine in a way

---

[49] Apple Computer's Newton device was the first entrant into the handheld computer market, and is a major focus of Jay Levinsohn and Martin Meyer's presentation, which follows Nusser's presentation.

that was more appropriate for the task that they were pursuing. And we really did this seat-of-the-pants; now that I've been working in this area a little bit longer, our developer did a pretty good job of thinking about how people work with this information. But I want to take this from a more systematic perspective.

So I'm going to start out talking about: what are geospatial data? They come in a lot of different formats, and they have a lot of issues about them that really affect how we set up our survey data collection systems. And I'll probably only have time to talk a little about the cognitive aspects; there's a whole other set of problems that have to do with computing infrastructure, because these data are very voluminous and require some adaptability in how you work with them. If you want to know something more about this research, go to this Web site. This is also funded by NSF, the Digital Government project. And my collaborators are both at Iowa State and [the University of California at Santa Barbara (UCSB)].

So, we're used to dealing with this sort of thing; we know a lot about how to phrase questions, we have these nice clean coded texts and— even if we're doing a scientific study—we like to develop protocols to construct precise and definable measurements. When we move into geospatial data, it's a different ball of wax. There are two general types of spatial information: one is called vector and the other called raster. Vector data are just points; lines are connections between points, and polygons are just a bunch of segments put together to form a polygon. The basic form is what you get out of your GPS unit, which is a single point, or a sequence of points to make a line. This is a road map that is a set of lines or polygons; if you tap on these lines it'll give you back the street name. So, even though you have this spatial information, there's sometimes attribute information linked behind it. Raster data we can basically think of as two-dimensional array, where the cells are basically pixels and—in this case—the value is a color that's provided on the screen. So all the information is visual; there's no extraction of numeric information or identification of features, so it's up to the human to figure out what's going on here. Something like this, which is a soils map, you might be able to—it's a raster map—you might be able to tap on it, and there is some attribute information connected to it, what kind of soil is there and what are the properties associated with the soil.

More often than not, in our world, we're going to want to combine a couple of different sources. This is a topo map with a sample unit boundary. And we might even have dynamic data, even in the form of video, which I'm not going to talk today. Or it might be in the form of—I don't know if you guys in the back can see this, but—getting GPS readings as you're driving along or walking along a route.

So let me talk a little bit about how we might use this type of information in three different phases of the survey process. This is just sort of a backdrop on what the issues are. When we're doing "planning" here, I'm thinking about a field representative who has been given a set of sample sites which they have to go to, so they're working with multiple destinations. They need to know what order they should visit these sites in and what should the route be? So the basic spatial information here is just street maps—annotated vectors, essentially—and the determination of routes along the vectors in that street map. And, mainly, we look at things at a large scale; we might want to zoom in on some smaller areas as part of the preparation process, but generally we're in the larger scale when we're doing this kind of thing. When we're actually navigating to a site, we're generally focused on one destination at a time, a sequence of single destinations; we might start off with an overview of where we need to go, but eventually we need to focus in to find out what the specific instructions are, in terms of how to get to a place. Interface design is going to depend somewhat on your mode; if you're in a car, you don't have hands or eyes particularly available for looking at a map. If you're on foot you have more availability. How quickly you go determines how fast that map might be going by, and that interacts with screen size. So there [are] some issues here we need to deal with.

Also, because we're mobile, we're starting to come into this dynamic component where we might have a GPS moving along in a map, and we get to the first data collection we want to do, which might be capturing a route to get back out of a convoluted neighborhood or for the next field rep to be using.

Finally, in data collection—this is really tough, and there's a lot more to be done on this side—it's usually a very focused activity at a specific location. And we go beyond our GPS and maps to using other kinds of information. If I'm doing a Census Bureau-type study, I might be having to go out and do some listing. I might want to be looking at plat maps, the map address file. I might be interested in topographic maps if I'm in a new neighborhood and really trying to figure out, "Where am I in this neighborhood?" And we're using this material as reference material, to find out what is going on around us in our surroundings or as the base for collecting data. So, if I'm collecting a GPS location I might want to see it on the backdrop of a photograph to make sure that I'm actually collecting the right thing. We might also want to make hand annotations on an image; I was actually just talking to Marty last night about doing just that sort of thing.

So what are the issues for survey data collection? Well, the first thing is that spatial data—and raster data in particular—can be quite large in size, which raises computing issues that come into play in terms of transferring data out and where you want to do your computing, whether you want to do it on the server side or out in the client. The other big thing is that we are very focused on error in the survey world, and all locations—all geospatial locations—have some error, and it may be acceptable and it may not. And part of it comes from the fact that we're trying to take the Earth's surface—an irregular sphere—and put it down onto a screen or a piece of paper. And that involves some distortion, and there are many different ways in which we can do that, so it's also possible to get geospatial data sets that have two different formats, put them together, and it's not a good idea. It's very easy to make mistakes due to the complexities involved in spatial data formats. There are also issues of resolution and the positional accuracy of nodes in your vector, and so on. But, once again, it's always going to be present; it's just a matter of whether it's going to be *too* much or whether you can live with that amount of error.

This gets back to some of the things Roger was talking about. Historically, the primary mode of interaction with geospatial data was visual. We have very little feature extraction associated with ... particularly image data, to help us as humans to extract information from some of the more complex formats. I really think that's down the road; it's not something that we can't do, it's just something we haven't focused on.

In addition, we have a whole new cognitive process we need to think about, and we need to find out whether there's a framework we can work from in presenting this information to the interviewers and perhaps someday to the respondents. So there's a literature on spatial cognition theory, and one of the very first questions we asked was, "does this literature apply in our world?" I'm going to talk a little bit about this, and one thing to keep in mind is that there's a lot of variability in how people perceive spatial information, so we want to bring that back into creating settings and interfaces and tools that allow us to get back to our principle of repeatability in data collection.

So the basic goals for using digital geospatial data in field data collection [are] to develop designs, tools, and systems that will minimize measurement error, in part by accommodating variations in the users, and to be able to provide appropriate information for a given field environment. I probably won't talk very much about this, but ... on the computing side, we're trying to develop systems that will allow us to have different kinds of computers and different types of abilities out in the field. And sort of make the infrastructure much more flexible than we're used to doing.

I'm going to talk a little bit about spatial cognition theory and interface design. And then talk about a user study I did with Jean Fox at [the Bureau of Labor Statistics (BLS)] that helped us get the first step under our belt in this area.

I'd like to talk about developing spatial knowledge, and I think about this as though you're moving to a new area, a new city. There are basically three levels of knowledge that are conceptualized in this theory. The first is that you have "landmark knowledge." You have isolated points on the ground that you are aware of, but maybe you haven't connected them. So it's the very first stage, and you pretty quickly get to the stage where you can connect lines between points. And this kind of knowledge is called "route-based" or "procedural knowledge;" you tend to be moving along in sort of a linear sequence in your mind, you tend to be orienting yourself on the ground, looking forward with left-right type reference frames. As you become more familiar with the area, you develop much more detailed knowledge. Your network begins to fill up so that you have a pretty detailed network from which to work. Your reference frame tends to switch to a "birds-eye" view where you tend to start viewing the system from above, with cardinal directions. And people who work in this framework usually work well with cognitive maps. How far along you get along in this development, and how quickly, is mediated somewhat by your spatial ability. The dimensions of spatial ability that are important for this problem are the ability to rotate information, so in your mind be able to rotate a map and see where you're going even if the map is north-up; to orient yourself or find yourself on a map; and then confidence in or anxiety over performing spatially-oriented tasks. And what appears to be known is that as we get older our brains don't work as well as they used to, our ability to rotate information is not as good, but we know more. We have more experience, and so are able to orient ourselves on maps more readily. An interesting aspect, it turns out, is that our anxiety over performing spatial tasks varies very widely across people and sometimes within people. So if you've been lost, and you've gotten sort of befuddled, and you have trouble placing yourself on the map, this is kind of your mental state decreasing your spatial ability, and that's important when you get lost out in the field and are trying to find your way to your sample site.

So, you can think about different strategies that people use in dealing with spatial information as being a continuum, from route-based thinkers on one side to configurational or map-based thinkers on the other. Route-based thinkers tend to think of things in linear order, an egocentric reference frame. They tend to prefer written directions, lists of instructions with landmarks noted on them. They tend to want to stay on those major thoroughfares because they haven't developed that network. Whereas people who use the map-based view tend to think from the bird's eye perspective; they're very comfortable with maps, often they have just a map in their head and don't really need a paper map or a digital map, and they're willing to get off the road to get to where they want to go.

OK, so let me shift gears a little bit and talk about how we work with maps and the navigation process. What you're doing is aligning three different reference frames: one is yourself, another is the reality around you, and the third is the representation of that reality, the map. If these things are not aligned, you have more mental rotation to go through. Most of the maps we're used to looking at are north-up maps; the nice thing about these is that, for example, if you're a pilot and you have to communicate with air traffic control you have a very clear communication mode with this kind of a system. But you have to do much more mental rotation, and that raises the cognitive load involved in using this sort of presentation. Research has been done to compare with "track-up," or "head-up," or "forward-up" type of orientation where the map is always turned in the direction that you are going. This eliminates the mental rotation step. That's good for our setting in the sense that, usually, interviewers are working by themselves and not needing to be in communication with someone else. When you add driving in, you've got two really demanding tasks, cognitively. So when we think about using spatial information in this setting, it's very important to try to minimize the mental effort needed to interact with the spatial materials.

When we think about map interfaces, we need to think about our task; I'm talking about navigation right now, but it might be a different task in the survey process. How someone prefers to get and think about their spatial information; which way is it easiest for them; and some issues related to presentation. You only want to have the stuff you really need on that map; it needs to be very visible, so if there are any cues to assist you in finding yourself on the map the interface design will be better. So, that's a quick primer. Any questions at this point, at all?

So what I'd like to do now is talk about a small user study that Jean Fox and I did. We basically used field staff that are associated with the Commodity and Services Survey; that's an establishment survey where they're going out to collect prices on goods and services that people have given them through another survey. They get new businesses that they need to go to through the refreshing of the sample or by being assigned to new areas, and that happens quite frequently. We were focusing not on the Wal-Marts and the McDonalds, but on the things that were harder to find—the service industries, gardening services, upholstery, medical services, and so on. And also looking at areas that they were not familiar with. And the questions that we were interested in answering dealt with these modes of dealing with spatial information, and how they impact survey work. And, then, is there evidence that if you give somebody a digital map versus a paper map that there's any bang-for-the-buck?

Third, if we then add a GPS to that environment, will that help them in any way or is that just gilding the lily. And we got pretty lucky; we worked with four staff, and they laid out very nicely along this continuum between very route-based and very configurational. There was one person who hadn't been very long in the city she was working in and had fairly low spatial ability; really preferred to work without a map at all but with directions listed down. At the other extreme was someone who had lived in the city for a very long time; she was willing to use maps but she had it all in her head, so you could tell her something and it would log in her head and she would be able to go. She didn't really need to look at a map much at all because she had it all in her head.

What we did was—thank you, Marty, again—used one of these. We also looked at using an IPaq which is more this size, but there were software issues there so that we didn't get much out of that.[50] We had a very cheap software, Streets and Trips, made by Microsoft—lots and lots of features, really great for this experiment. After training them, we gave them 6 to 8 outlets and they had to plan their route using the software. They had to navigate to half of them by only using the digital map and to the other half adding in the GPS. The way the software works is that you put your businesses—you can use Excel for this—or other addresses into the Route Planner; you can ask it to optimize the order, given the beginning and end of the route; and produce directions in both written and visual form. The staff did that; they generally accepted the order and they always accepted the route. So you're left with dealing with a map and route listing interface; you can adjust the size of these, you can adjust the font size, you can keep or hide features like hotels and restaurants. There are a lot of things they could do to alter the interface.

What did we find out from this? The best thing we found from this was that they mapped very nicely into the spatial cognition framework. The behaviors that they exhibited followed closely what would be expected from the literature. People who are route-based tend to make that part of the interface much bigger than the actual map; vice versa for those who are accustomed to looking at maps. One woman who had pretty good spatial ability but was new to the area would use her map as a set of instructions; she would zoom in and kind of scroll through, so she was soaking up the step-by-step instructions but in a visual form.

The preferences for interfaces, and this is a fairly loose part of the study—they basically didn't do certain things. They did increase the font to make it larger, so that they could see more readily on the screen. We told them about the landmarks they could add to the screen but they never wanted to do that. They were most interested in cross streets

---

[50]The IPaq is Compaq's handheld computing device.

and the street number of the business they were headed to, and they had all sorts of suggestions to try to make that more visible. They really preferred the simpler interface and did not try to clutter it up with anything that seemed like marginal info. GPS was really quite effective at reducing the cognitive load associated with finding yourself on the map. So, what people could do is that they could pick up the tablet and see whether their icon—a little green icon—was on the line or not. Just a very quick assessment, so they could take a quick look and keep driving. If they got lost or on a different route, they could pull over and immediately see that, well, I'm here, and this is where I'm supposed to be and they could choose a new route. The software could actually help them with that, but we did not teach them that feature. The people who used this feature were really much more comfortable in doing this assignment because they got this quick and easy feedback; the route-based person, of course, didn't use this because she didn't look at the map much. If you're not using a map, it doesn't do you much good to do GPS.

So our conclusions from this study is that it's probably possible to identify a couple of different approaches to developing interface design, or to make the interface so that it's suitable to a spatial strategy. It's possible for the setting of personalized training, to take someone who is route-based and try to move them towards being more map-based. We thought that this would be very desirable to get people up to the level of survey knowledge [before getting into] the navigation phase.

There are a bunch of unanswered questions. The first ones on our docket—we're just in the middle of planning a second study—include: what would happen if we left these tablets with them for more than a couple of days, which is basically all that we worked with them on this time. Would they change how they use the map resources? Would they be more flexible with them? Would having a GPS help those route-based staff use maps more readily? Would it provide those additional cues that would get them over the hump in using maps? And then there's a whole number of things that GPS software *can* do—not the one we were using—in terms of recording routes, capturing coordinates, and using voice cues in navigation. So there's a whole other set of interface questions looming on the horizon, if we use another piece of software.

What we haven't dealt with are a couple of very big issues. One is raster data, these more complex images. Spatial cognition theory is really centered on way-finding and maps; what do we do with things like this? Is there a way in which we can augment it so that laypeople can be using this kind of interface, for example, if we were going to be having them doing decennial census work? The other thing that we really haven't broached at all is: what principles should we be thinking about in *collecting* spatial data? We've got a little bit of information on that on the point level from our other studies with USDA, but there are a lot of questions there.

The second big area—and an area that's of interest to me, and that Marty's going to spend some time talking about—is what can we do with small interfaces. So, we're very interested in using something this size [*holding up a Palm Pilot/IPaq-sized device*] because the tablet—particularly for environmental surveys—is just a little bit too big for the things that people need to do. There's also—part of the project is looking at other emerging technologies. This is a set of glasses where the screen is being projected into a little inset in the glasses, and it would actually overlay information on your view. So you might have a see-through view. And there are other questions related to other modalities coming down the pike.

So, this is really the "What are we going to be doing in five to ten years?" type of research. So let me just give you a little bit of a heads-up of what we're looking at in this project. This is actually a different kind of interface where someone has a screen clipped to their glasses, so instead of being an overlay like this it's really more like a separate screen off in your peripheral vision. People at UCSB are looking at how you might design that interface and how you would interact with that interface given that it's up here instead of in front of you in a tablet. They're also hoping to explore—although this is probably next generation stuff—how to use a see-through screen; in other words, if you overlay screen information on the environment, can you label things like houses and say, "this is the one that you want to get to"? And how much dissonance is there when you have this in your vision field and you're trying to do other things?

Where am I on time?

CORK: You have 10–12 minutes.

NUSSER: OK. I can pause here for questions. Yes?

ROBINSON: You said your route-based person didn't use maps . . .

NUSSER: Right . . .

ROBINSON: So what did that person do when she got lost?

NUSSER: Well, she had to use the maps to recover. But what she really did was—the map software would through this particular route. She would look at the map to begin with and pick the interstate—to go 15 miles out of her way to do something that was very comfortable. So she was less likely to get lost but she was spending a lot more time getting where she needed to go. So that was an overstatement on my part.

GROVES: A similar sort of question: How did you sort them into these headings? How did you measure their spatial ability at day one?

NUSSER: We were not permitted to do testing on these employees, so we did this through observation—looked at how they were using the maps, setting up their screens, and so on. So that was a post hoc classification of these individuals.

GROVES: And how did you know how they performed on the task?

NUSSER: We were with them in the car. And actually one of the questions we have for next time is how can we get that assessment a priori. We have some—I think people are pretty good at rating their spatial ability and their preferences, and so we need to come up with a set of questions that will allow us to do that ahead of time.

CORK: Any other questions for Sarah?

NUSSER: Do you want me to talk about computing stuff at all?

CORK: You have time, so go for it. There's one more question ...

PARTICIPANT: You said that many people, I believe, were interested in the address, and had suggested changes, and I wonder if you could explain a little.

NUSSER: Yes. Their job is to navigate to a housing unit or a business, and the key identifier there—once you get to the right street—is the street number. So they're driving, and they know they're on the right street; then they need to know that they're on the right block, and then what the sequence of the street numbers is. And they wanted that street number to be about *this* big on the screen because we were in larger cities, and the street numbers could be four or five digits long, and you can't remember that.

PARTICIPANT: [*inaudible, but centered on differences in spatial ability between right-handed and left-handed people*]

NUSSER: I'm trying to think ... yeah, that's right, and I think we had all right-handers.

PARTICIPANT: Because they think very differently ...

NUSSER: I think we did have all right handers ...

PARTICIPANT: [*inaudible*]

NUSSER: I'm married to one of those. [*laughter*] But I think we had all right-handers, all female right-handers. There's also some literature that says that females think differently from males spatially; there's also in the literature a convergence in how they think, so I think that my personal theory is just a little bit like the Web: as time goes on, and socialization becomes different, then people will tend to be thinking in the same way.

MARKOSIAN: I didn't quite understand how you could differentiate preferences for north-up versus head-up maps ...

NUSSER: We could not, because this software only does north-up. You would need a compass to do that, and we didn't do that. The GPS we had can't do that.

MARKOSIAN: The newer GPS handhelds can provide that.

NUSSER: Right. Jay?

LEVINSOHN: Do you have a sense of the trainability on these, on how possible it is to train people to modify their ability to use the more advanced capabilities? ... How likely are these devices to move people forward?

NUSSER: Yes, I—I don't think we have a ... The question was, do we have a feel for whether it would be possible to move these route-based people forward—well, not forward—into configurational points-of-view. I don't have a very good feel for that from the experiment. I do have a theory that if we add cues to that environment, that they can hang on to—"you are here"—and use to orient themselves—"you want to go here, and the path is this," that would be much better than just getting a map. Their prior experience is just to get a paper map that has no annotation that's relevant to them. So that's one thing we'd like to test in the next study.

LEVINSOHN: Give them personal icons ...

NUSSER: That's a very good idea, actually. Do we have time, still?

CORK: If you want to take a few minutes, sure.

NUSSER: I'll just go over this quickly; I threw this in because we had this comp sci community, and in the survey world we tend to have a very rigid structure in the way we set up our data collection systems. So I'll just throw these ideas out here. We tend to work in a system where we've got people out in the field, and we've got a repository that we've prepared, and a client-server interaction where we've prescribed everything that's going on between these actors. And part of the research project is looking at a different kind of model, now that there's so much available on the Web—and that's particularly true of spatial information—where a field user could not only get to the stuff that they need from a methodological point of view but other information that might help them when they're in a situation that's not covered by the repository prepared for them. And we'd like to do that in a way where the user can be completely naive; it's mediated by the infrastructure. And so the goal is to seamlessly deliver spatial or other kinds of information in formats appropriate to the field environment. "Appropriate" is a very loaded word; here, we're thinking about variation in user characteristics, variation in what kind of field computing environment we have—whether we have any number of the devices you're going to see in the next talk, and variation over time within a survey or a task on how these characteristics change.

So, just to give you a better feel for that—again, we're thinking more broadly than just demographic surveys in this research project. People come in with different levels of spatial knowledge into the project—you might have a social scientist who works with maps all the time, who will have a higher level of spatial ability than somebody like an interviewer who uses maps for navigation but their primary duty is to interview somebody. And then, working with the decennial 2010 group that's thinking about mobile computing devices, they're basically assuming that their workforce is citizens—we make no assumptions about baseline ability in terms of spatial strategies. We have the spatial strategy component that I talked about earlier, and then physical aspects having to do with how quickly you're travelling, whether you're travelling by car or foot, what you're doing—do you have your hands and eyes free, and then disabilities that really are no different from these physical settings, just an additional cause for having limitations. In the field computing environment, we're sort of thinking about the traditional things: what is the screen size, what is its resolution, is it color, do you have a little device that doesn't have much storage or processing ability, and how much does storage decline over time as you collect information, communications aspects and interface modalities.

So the notion behind this project is to borrow from the notion of a "data wrapper," which basically provides an interface between the computing infrastructure and the data. So you have a request coming into a front-end server that is shunted to the database; well, this wrapper will translate from—if you'll allow me the colloquialisms—the language of the infrastructure to get a query into the database that the database understands, and then generate data back that the infrastructure can understand. So, in this project, we're developing the notion of a "field wrapper" which is basically the same sort of thing. You have different conditions associated with the device, your communications, and the user. That's metadata about the field environment; that's shunted into the infrastructure through an interface, if you will, that translates into the language of the infrastructure.

So the idea would be to send information on this field environment forward to what's called a mediator that generates a sequence of actions that are appropriate for this interface. So in this case there may not be too much processing capability, not too much storage space, so the sequence of actions would be: get me a big chunk of data from here, put it in the computation space, and get ready to send out little chunks, several little chunks at a time, as you move around in the space. The upper setting, that's basically a PC. So you might ask it to go ahead and get a chunk of data here, and it might be wired a different way via a land-line or something like that. And you can just go ahead and transfer the entire file to that setting. So the idea is to get us away from this notion that we always have to buy 500 of the same box, and move into more of an incremental change strategy as was talked about yesterday, of having some like this and then—as new products come on—be able to add in different kinds of devices out in the field.

So I'm probably by now way over my time, but let me just say: there are a lot of problems here. And why are we doing this? Well, almost everything is tied to a point on the ground. This is becoming an increasingly common mode of communicating: we have our MapStats up for the federal statistical community. If you take the "geo" out of what we've been talking about today, you basically get the flowcharts that were up yesterday. So we're getting very used to interacting with visual information. And geospatial data are really rich in information content—much more so than a list of addresses—and I think that there's a lot to take advantage of there. And the quality of the data is increasing, so we can use it in a way that will feed into our measurement error structures. So, that's about it.

CORK: We can take one or two questions for Sarah while we switch computers up here.

PARTICIPANT: How large are the files? You said that they are big . . .

NUSSER: They range from just a few kilobytes—which is what a vector file would be—into megabytes. Generally you're not dealing in gigabytes, but it's not hard to get there fast with a few images. So this presentation which had a number of clips in it is probably 10 megabytes. When we think about the data for the National Resource Survey which we do—which has about 300,000 sampling units—we're definitely thinking terabytes for that data.

PARTICIPANT: So that is something you really need to think about.

NUSSER: You really do, and I guess in part that's why we're not thinking so much about data structures but rather how can you work with—for example—storing this in different places and being able to pull it into a computation server to break it up and send it into the field, rather than having things set up in a prescribed way before you go out in the field.

PARTICIPANT: [*inaudible*] So would you need some kind of dial-up connection?

NUSSER: It can be slow, and so if you're doing dial-up, this is one of the settings we're working with in the project, where you have pretty poor communication. So you want to take just a subset of the data and send it out, something that's small, that you can send out via a wireless or dial-up line, or whatever. And we have a long ways to go. Yes, Mike?

COHEN: Handheld devices are being trumpeted for use in data collection for the 2010 census. Do you see the need for different types of devices inside the blue line versus outside the blue line?

NUSSER: Sorry, could you give me the jargon?

COHEN: Sorry; basically, urban versus rural.

NUSSER: Well, in the urban areas you're more likely to be using maps. And in rural areas you're more likely to need to rely on photographs. But I still maintain that there's still a lot of overlap; one of the areas that we're proposing to focus on is new developments. You have—TIGER isn't going to have lines in there for the new roads, necessarily. And your urban datasets on the streets may or may not have those; you may have a recent photograph, however, and you might want the enumerator or some staff member to record what those lines are, either by GPS or by annotation. Harry?

ROBINSON: Partly a comment or a perspective ... your dichotomy of route-based versus map-based maps very well onto the different ways that people use and test applications. When people aren't familiar with an application, we give them a tutorial and they work step-by-step through it. The same way when they test: they record step-by-step. And as they get more cognizant they get more of a map-like view.

NUSSER: Yes; I think actually if you give interviewers instruments, it's the same thing as well. Their thinking, over time and with experience, becomes more of a navigation view in mind. Yes, Miron?

STRAF: If you can track across time, have you looked at any effects of climate change?

NUSSER: No, we haven't looked at that at all; that's a big area.

CORK: Thank you, Sarah.

PROSPECTS FOR SURVEY DATA COLLECTION USING PEN-BASED COMPUTERS

Jay Levinsohn and Martin Meyer

CORK: Continuing and building on some of the themes addressed in the last talk, we invited Marty Meyer and Jay Levinsohn to talk about prospects for data collection using handheld portable pen-based computers. Jay is the manager of technology issues and Marty is a research programmer and analyst at RTI, and—apparently, in the tag-team situation here—Jay will be doing the presenting today.

LEVINSOHN: Good morning. We're going to talk about ... we've been talking mostly, the past couple of days, about things to put on devices. We're at the other end of this: after you've done all this development and after you've designed, what are the tools you can take in the field and what equipment can you use to display and to show the items you've developed? And, in sort of a global sense, what role, if any, do these newer and smaller devices play in survey data collection? Can handhelds be used effectively in the field for survey data?

And what's out there. Below that, what are the software design and data transmission issues that need to be considered if you're thinking about this equipment?

It's relatively early. You know, handheld devices have been out for a while; you've seen some of the antiques being used. The Newton devices have held, I think, their ninth birthday this year. So there have been devices out there for that long, but their appearance in the market, their volume and their pricing, have begun to change very, very rapidly in the past 12 months. The Newton was a $1,000 device when it came out, and there is pretty much nothing in that price range for these small handhelds now.

So, we see the use of the handheld whenever the full laptop is not an option. There are certainly differences in capability and form factor for these devices, and I think you'd want to make the decision to use a handheld where it's most effective. Clearly, for doorway screening, where you need a light, portable device with a good screen; hospitals and medical clinics, where you're moving around and very mobile, going from room to room collecting data. Another example is diary applications, where people are recording information through the day. There have been studies, EPA studies, where you want to report your environment all day to assess your exposure to chemicals. Or, if you're working in a plant, how many hours did you spend in this room versus that room? How much were you exposed? These are easy devices to carry around to do that. In market research, people tracking you down in the mall and getting to you. Fundamentally, any situation where the data collector is on the move, needs to be standing during the interview, and needs something small and light.

So far, it looks like you would need to restrict your attention to shorter and simpler questionnaires. Some of the—Moore's Law will probably have some impact. These machines are expanding in capability fairly rapidly, certainly in data storage capacity, and the ability to plug things into them to expand their capability is improving very rapidly. I haven't seen so much change in the speed of these devices as you see in other machines, and that may be related to the effort to keep battery consumption down and weight down. And maybe because the market is still relatively small. But I think that may change and you may start to see this doubling of capacity every 12 to 18 months.

Based on our experience at RTI, we've been using as a handheld the Apple Newton which—in many ways—is a very innovative and very clever device. It's small; it's light, about a pound and a half. It has a large screen, by today's handheld screen size standards. And it's been around for a long time. We're using this in the National Survey on Drug Use and Health, and we've had it in the field for three and a half years. So we have collected a large amount of in-field experience with this device—over two million hours of use—and it's been positive. The device has held up; the interviewers have learned how to use it; we can maintain it. Even though it's out of production, and the software is frozen, it's still a working device. Apple is still maintaining the device. So our experience with it has been good. We're doing fairly complicated screening with it; sample selection is built into the screening algorithm. And we're using it nationally, in every state, so we have used it in every climatic condition you can think of: freezing in Alaska in the winter, summers in DC and New Orleans, and we have people in Hawaii. So it's done very well. Probably, its utility and repair rate is half that of a laptop. It has no moving parts; it seems to take a beating fairly well. We bought 1,700 of these devices, and over a four-year span have lost about 100 due to getting run over by baggage trucks, being left on top of the car. But, in general, they're doing very well.

So we'd say that the answer to "for this class of device, is it useful in the field?" is yes. We've been happy with those results.

We're also beginning the search for a replacement for this product. The Newton, and the Newton developers, have moved on to Palm. They've moved from Apple to start the Palm [platform], they've moved to Mindspring. This is sort of the seed device in this market, and the products today are still vaguely reminiscent—there's a Newton flavor in all of them. More so in the Palm class, but certainly in the others. So we're looking for a more modern device that's faster, that has more capacity; the memory capability on the Newton is somewhat limited, and the speed is slow. What we've found is that we've filled the Newton in terms of software capability, and speed has been an issue for some of our interviewers who have a large workload. Some percentage of our interviewers, when they get up to a large number of cases, it begins to bog the Newton down more than we would like. Plus, it's not in production [anymore], so we need to look at new devices.

The questions are, do we want one ... Currently, on the drug survey, we're using two devices. We do doorway screening with the Newton; we go in the house and do interviews with a regular production laptop. Do we want to look at that model, or do we want to move toward a single device?

So, we've set up some target requirements for evaluating these devices: what kind of device do you want for doorway screening or a very mobile activity? What kind of device do you want to do an in-person interview with a pretty broad range of respondents? Our age range is from young to old, good vision to bad vision, good reading skills to bad reading skills. So the population for the interview is pretty broad, and it can be a pretty lengthy interview—anywhere from 20 to 70 minutes. So do you want one device to do that, or do you look at two?

These are sort of a list of things we looked at for this application; perhaps the column on the left is a more general thing to look at. These are things you need to be concerned about in evaluating a device. Battery life, screen readability, can you hold it in one hand. Durability—will it hold up? What happens if you drop it? What happens if it gets wet? What happens when it's 105°? Most of these devices have problems above 100° and below 20°. And as you go further away from those extremes they begin to stop. At 110°, 115° the LCD screens tend to ... you can't see them. They turn black. Or, when it gets to 5° and the wind's blowing in Alaska, the screens again lose their ability to display. You're not going to be doing much doorway screening in those conditions, though. [*laughter*] They're working at it ...

PARTICIPANT: Do the interviewers tend to get frozen?

LEVINSOHN: The interviewers give out first, that's true. You know, I think there's a human effect: if the people can stand it outside, the devices will typically work.

There are also things you need to evaluate, in how it looks under indoor and outdoor light, and how it deals with very strong direct light. If you're in a very sunny environment, some of the LCD screens don't do that very well. And you want to look at what communications tools are there, how much memory can you get in it. We feel like there should be permanent storage; need permanent storage in these devices. Some of the Palms, some of the other devices optionally have a permanent storage. On the simple Palm device, if the power goes off, you've lost all the things in it, unless you have a permanent storage device. But most of them now have a slot you can plug something in to, and the things that go in there are getting bigger every week. And cheaper.

Vendor support is an issue. We've been in the field with a discontinued item since we started. It's worked, but it's a little scary. Nothing can change, so if there had been sort of a sea change in transmission protocols or maintenance procedures, it would have left us at risk at some point. It's a little hard to guess; even big players can drop an item. So the market is pretty volatile in these devices, and I think it's hard to guess. So it may not be a given vendor—it may not be Panasonic you're worried about—but [are] there at least going to be a couple of vendors in this market providing alternate equipment and supporting protocols, software, replacement parts, and hardware?

Operating systems—you know, you may not have a lot of choices with a handheld device. They tend to be more esoteric operating systems, and you're going to end up programming in that environment. In the laptop environment, you might be looking at a Windows environment—and if you're looking at Blaise or standard questionnaire development package—you might have to have a Windows 98/2000/XP-type operating system.

So what's the landscape for these devices? There are Palm-class devices which—now—there's a number of brands. They're this size form factor, 3 inch screen. They come in black-and-white or color, and the displays are pretty good and pretty readable. They are small; extremely light. Depending on the models, very, very good battery life; some of these will, with a couple of AA batteries, last days or weeks. The newer models are rechargeable; they're certainly good for a day's work. They have good capability. They tend to be relatively slow and have small amounts of memory, but you can plug in cards with the newer Palms. And there's a good mix of manufacturers, several people; Palm makes these, Mindspring makes these. There's a new vendor that pops up all the time.

The next class is something called a Pocket PC. This is, in some ways ... there are emerging standards for these devices. They're bigger in terms of screen size; they're bigger in terms of processor speed. There is a significant difference in the processor speed. The 200 Mhz are the top end of these devices; some of the earlier Pocket PCs are slower. They have more memory; they have a different operating system, and the screen is significantly bigger. And they also come in color or black-and-white.

We talked about clamshell or hinged handheld PCs. We don't have one of those to display, but by clamshell we mean a device that folds in half and the screen would be the top half and the keyboard the bottom half. Sort of like a PC, this format, but much smaller; maybe a half or a third of a PC. A lot of those run on CE; you can also find them for Windows.

Then there are Windows CE tablets, and this is an example of a Fujitsu tablet. This is an older-model Fujitsu. It has a *bigger* screen, so now you get to see a significantly bigger screen size. It would have the same computing capabilities as some of these devices, but it has a nice large screen.

Then there are what we call Windows tablets, which are bigger machines. This is an example of a Fujitsu tablet that has a somewhat smaller screen size. And this is the big Fujitsu that has the biggest screen. All these differ in weight; we have a dramatic change in weight and size and portability and battery life between these two formats. But you have a *huge* screen difference. This one is a Windows 2000 capable machine; they're reasonably fast computers, 400 to 800 MHz machines so that they're beginning to rival what's on a desktop; an 800 MHz machine is a pretty current desktop PC. They have large amounts of internal memory, RAM, and they have a hard disk. So a big change ... the CE class machines generally do not have a hard disk, they're using some form of volatile or non-volatile memory internally. So that might have some impact on your ability.

Then there are sub-notebooks and miniature laptops, which are just small laptops. They tend to be, oh, they vary in size, but they're typically a laptop configuration and they're hinged.

Finally, this last class is relatively new, and still the form is emerging. The standards, it's not clear what they will be. These are smart cell phones—cell phones built on top of a [personal digital assistant (PDA)]. So you might have a cell phone sitting on top, or inside, a PDA; it might be a Palm operating system. Not only do you have a phone built-in, you could also have a Palm OS built in. There would be messaging capabilities and abilities to send and receive built into the phone, and capability to program and connect those two devices together. There's a class of wireless e-mail devices, like the Blackberry and some others, that might be appropriate for small data collection formats. I'm not sure that any of the federal-size surveys we've talked about here could use these smaller devices, but they're interesting and they're changing quite quickly.

Of course, wait 'til next month [*laughter*] and this list will be longer. The handheld of this is now the volatile part of the marketplace. A few years ago, you saw this kind of development in laptops, where every week someone was coming out with a bigger screen size or larger hard disk. They'd finally get high-end Pentium processors into the laptop and, every month, the manufacturers were beating themselves to the market with another device. That's exactly what's happening now. So, all these devices that we bought—I don't think you can buy any of them now, and we just bought these six or seven months ago. All new models; I guess the Fujitsu is still the same. But the Pancentra is really not on their main market list. This is their newest device, the sort of mid-size screen; I think this is a Pentium 600, so it's a pretty high-end computer, and a small—very good—screen. The Compaq and all the Palms we bought, the Pocket PC models we bought, have all been leapfrogged by other offerings, both from the same vendors and from other vendors.

It's very rapid, and—if you're going to pick a device—you're going to have to take a guess. I think you can pick an area, like the Pocket PC, and assume that that environment and those standards are going to be there, but the device you do your development on is not the one you're going to be going into the field with. And it may not even be close, if you've got a long development cycle.

There are a class of design and performance trade-offs that you have to evaluate based on the application you're looking at. The different screen types ... the screens come in different types. There are some that are optimized for indoor use, that are backlit, that will work in a dark room, look real good inside—and disappear when you go outside. There are outdoor screens that are just the opposite; they work very well under bright light that's reflected off of the screen and are very good, but go into a dim hallway and they're not very good. And then there are screens that are built with some backlighting or edge-lighting that try to do both. So you have to evaluate that in terms of the application going on. If you do a lot of outside work, that's more important; if you're all inside, you may want the best indoor screen you can get. Display size and resolution is really a form factor issue; there are surprising standards within the devices. No one has really come up with a Pocket PC that has a really super 1280×1024 resolution; they're sticking within that framework. Permanent storage capability, capacity, are trade-offs between these devices. Battery life, clearly, is a big difference; the battery life on these devices with a hard drive, with a big screen, with backlighting, are going to be nothing compared to smaller devices like the Palm.

So you may need multiple batteries, you may need battery management. The manufacturers ... that may be today's big lie, how long the batteries last. After "the check's in the mail," the "four-hour battery life" may be the next big lie. You can get four hours, maybe, but it depends very much on what you're doing, and if you're using the device continuously. On the health survey that we're running, we have .WAV files and a lot of disk access that may go for an hour and 10, 15 minutes. And that's a pretty consumptive activity; we're spinning everything in the machine, and in the laptops we have, the battery lives don't last very long. So it's a question you want to evaluate and if it's not going to meet your needs, you have the issue of managing multiple batteries— buying batteries, which are expensive, and providing some tool for your interviewers to charge them. You might have to buy one of these little "toasters" for them to put the batteries in and charge them up at night, or they could get car adapters and charge them in the field if they can do that, or they can go to lunch and find some place where they can plug up.

Durability is an issue. My suspicion is that these devices will be like the Newton; there [are] no moving parts, they're relatively light, and they'll probably hold up pretty well. And there is a lot of experience in the field with these devices, you know, as a commercial application. As you move up in screen size, get a little more parts on it and a few more cards in it, it may get less durable. With the bigger devices, like this Fujitsu, if we drop it, I don't know. [*laughter*] And I'm not going to do it. We couldn't get the Fujitsu engineers who came to do this test for us, either. They're reasonably well-constructed, but they have a lot of parts in them and the screens are fragile. The screens are big and if you drop it on a corner, I think it's history. They're relatively expensive devices; this is well over $2,000.

Size and weight are an issue; transmission capabilities—what devices are on there and what speeds will they run at. Software and questionnaire development tools are all things that need to be balanced in picking a device.

This table sort of let us put these things together in one place, and one of the questions we had in evaluating a new device for the National Survey on Drug Use and Health was: should we stick with a handheld device for screening and a laptop for indoors, or should we try to do all of it with this device [*a tablet*]? Go to the doorway, screen, go indoors, set this up on a stand with a keyboard and turn it into the equivalent of a laptop. There would be a lot of advantage going back to one device; go to a stronger operating system, go to one device to cut down inventory. So we're trying to evaluate that. This table sort of got at our sense, or initial thoughts, about whether there was one device that could do both things. Basically, we said that it's only the tablet. We certainly couldn't do the interview . . . we have 2 gigabytes of .WAV files that we play for the sound parts of the questionnaire, and it eliminates most things. If we only do one device, it would have to be a tablet. But the Pocket PC, we felt, would be viable for our screening operation; the Palm was a little too slow and a little small and—at the price differential—wasn't enough to get too worked up over. If cost is a very big issue, then you're looking at Palm devices. To get below $200 per unit, you're looking at Palm.

So we felt that the Pocket PC would be fine for the screening operation and a likely replacement for the Newton—better in many ways. And we felt that the tablet would be a good device for doing both, if we needed to, with compromises. This is a little heavy to be at the doorway but, still, it would work. And we're about to start focus group testing with our interview staff and with some elderly people to see how this screen works for visibility for doing an interview. And that's going to be done by the end of the month.

Well, once you've made some decisions about the physical parameters of the machines, what are you going to do for development? The development tools are pretty good for these devices, and dramatically different than when the Newton came out. The Newton had a very limited set of development tools and a small developer community, and all of those guys are doing something else now. There are pretty good development tools for the Palm OS environment; the basic units are where a lot of development work is being done. There's CodeWarrior, which is a C-based tool. There are some free tools; there are a lot of other tools for the Palm environment. There are rapid application development tools;

there is a package called AppForge, which plugs into Microsoft Visual Studio. It's very usable. There's a Canadian company which makes something called MSBasic that's easy for development. And there are lots of third-party tools. EntryWare is something I stumbled across recently, which is a questionnaire development tool which will allow you to author questionnaires and run them. All of these tools are, you know, emerging fairly rapidly—seem to be adequate for the task you want to do.

I'd say that there's more software, more development tools, in the Palm environment than in the Pocket PC platform right now, but that will probably change. There are tools from Microsoft; there is a Microsoft standard for the Pocket PC, Microsoft is backing this device and moving in this area, and providing good support tools for it. So there are synchronization software and development tools, so that the Microsoft embedded tools ... There are currently embedded C and embedded Visual Basic tools, and they will be built into the emerging Microsoft .net software. AppForge also makes a tool for the Pocket PC, and there's a third-party Visual Basic compiler and other people doing it. There are some relational database packages out there and tools to get at, for example, SQL server or other tools on a network. And there are some Java and Sun tools available, as well. We've been doing prototyping using embedded Visual Basic, and some of those tools.

The C++ environment might be the heavier-duty tools that you're really concerned about performance or doing real-time applications, or heavy number-crunching, where you've got to squeeze things out, squeeze performance. You'll have somewhat more options and more control. The Visual Basic environment has less control and is less integrated into the system APIs and the other applications. That may change over time; people are beginning to do a lot of development for these devices, and I think it's a *big* commercial market for software as these devices spread around.

This is a second page of different applications that could be done through these two tools.

Tablet PC development tools are the same tools you're familiar with in desktop applications. Tablet PCs are running [Windows] NT, 2000, or XP—all run and do pretty well on those. If you move to the Tablet PC you have fewer limitations and less learning curve; developing for the Pocket PC or the Palm, you need emulators or you need the devices, you need to be able to move the software to the devices. You need to worry about—typically, for development on the Palm or the Pocket PC, you do your development on a PC, you compile your code, select a target device, and move it over. Or you have an emulator device that runs on your PC where you can emulate, do quick debugging, flash it up to see what it's going to look like—you still need to port it over to the device. The environments to do that are OK, but it's a little more complicated than working directly on a PC. And you need to target ... most of these devices have different chips in them, so there's a different target set if you're working for the Cassiopeia or the Compaq, or this one has a different RISC chip in it, and the compiler needs to be told, "Where am I going? How do you want me to compile this software?"

Desktop software ports easily to Tablet PCs, and it's easy to move it. You can provide ... pop-up keyboard stuff is easy to do, and you have the availability of these add-on wireless keyboards. One of the things that comes with these new devices is a set of new capabilities that has not been in the traditional laptop environment: they all have touch screens. That's sort of one of the requirements in what we've collected, that they all have touch screens. They have handwriting recognition that's not bad; may not be ready for formal data collection if you care ... if small error rates are a problem, handwriting recognition still has a way to go. It might be great for signatures, for annotations, for things where the context will allow you to get the right answer. Single error mistakes in an address or a ZIP code or a Social Security number are a killer. But it allows you to get signatures and other graphic information that's pretty good. You can add cameras; most of these devices plug in for video stuff. The new ViewSonic tablet that's just been released has a built-in video camera, so you can do that. Global positioning hardware we've seen connected, and software is being developed for this that allows that activity. And they're all either coming with built-in wireless or are easily adaptable for wireless, and the varieties of wireless are expanding pretty rapidly. So these capabilities are either new because they're just getting out there or because of the form factor.

We wanted to talk briefly about options, now, for how you put these things together and how you connect your equipment in the field. This is also an evaluation that you need to do. Most of these devices are thought of as assistants to a desktop PC; for the commercial market, most of them are talking about applications where they're going to be synchronized with a desktop PC. You sync your mail, you sync your schedule. They expect you to be attached to a PC and doing updates periodically. That's a model you can use, so that you can transmit—and they provide good software that's pretty good for those activities and also has outlets if you want to do your own file movement or data transmission. You can talk to a laptop and have your laptop talk—through traditional channels—back to your corporate LAN or other computer networks you're working with. You can use the host PC to communicate with FTP or SQL server databases. That implies that you have a host PC around. Some applications, the handheld might be the only device they have, so this model may or may not work. Using the canned sync functions, and the ActiveSync or HotSync software that comes with the packages, puts you in a tighter framework for development; you might want to develop your own software for that kind of communication.

Another scenario is that you upload data to the desktop and that you're still doing your communication with dial-up. This communication can be done by wireless; there are infrared ports on these, and I think that all of these have infrared ports. This works for syncing, and it's not a bad method. It's built-in. You can use BlueTooth or 802.11.b Internet for high-speed links between these machines; you can do dial-up back to the LAN.

This third scenario is sort of a direct connection between handhelds, and these could be dial-up; this could be a network, so that if you were in a clinic and had wireless Ethernet in the building, you could be doing this. And you can have different communications, either to communications gateways or the corporate network. Pocket PCs support dial-up networking; like the desktop PC, [they have] RAS connections that allow you to dial-in and make connections. You can do e-mail, you can do data transfers. All of these are sort of another level of effort than doing it on a desktop; the desktop is older, it's been all worked out, and there are lots of models. This stuff is all very new, and there are new utilities to support it. There's clearly a learning curve and another level of effort to get this level of communication software in place. I wouldn't say it's canned; we're beginning to look at that to see which model ... if we went to a single device, we would be doing this from a Windows PC where we know how to do it. If we go to two devices, our decision is [whether] we're going to have the handheld communicate to a laptop and let the laptop do the communications back to the company. Or do we want these indirectly. Right now, we're using dial-up communications with the Newton, and our field force calls in practically every night from this machine, using 57 KB modems. It's been very reliable, I'd say, for dial-up communications, which [still] means that you have lots of recalls and the line gets dropped depending on where you are—Iowa, it seems, is not a good place to do dial-up, must be the town our interviewer lives in. Over the national network, dial-up is not the same. And certainly things like wireless, cable modems, and DSL don't have that kind of availability. But that model has worked very well with that device, and that model is available with the newer devices; they all are able to have a modem connected to them and do dial-up activities.

We're looking at things like doing more complicated access, where they would be doing SQL server transactions, getting access to and from the database. We're not thinking about doing this in real-time with wireless; I think in time that will be an option, and depending on the geography it's an option now; you can do cell-phone-like connections in some cities, you can do wireless Ethernet in other cities. But you can't nationally; there's no national wireless network that would be available to a national field force. It will come, but I think that the rate of expansion there will be slower; ability to get cable modem and DSL—two products that are pretty well understood, people know how to do them—is still very erratic, even within counties. It's all market-driven; if there aren't enough people there, you aren't going to get it. Marty lives 6 miles from me; he can't get DSL, he can't get cable modem, he's got a satellite. I can get either one because I'm inside the city limits, and the more rural you are the worse it gets. So I think those issues will cloud the wireless area for a while. I'm not confident about predicting where that will turn out; the rollout has been slower than I thought it would be. I thought it would be easier to get cable modem or DSL because I think it would be a good way to make a fortune for the phone companies. But they don't think that way yet.

That brings us to this screen on future trends. OS vendors are driving the technology trends; they are developing standards for these, and that's a good thing. There are hardware and software standards, and ergonomic standards, that are beginning to get there and become well-defined. Microsoft has taken the lead on these, where they've defined a standard for the Tablet PC and the Pocket PC. They're talking about a smart phone standard. The Palm is well-developed, fairly mature, stable platform that's been around, but the Pocket PC is giving them a lot of competition. If I were making projections, they're beginning to eat the market they had. So that there's going to be a lot of competition, and I see the Pocket PC as a very formidable competitor to the Palms, which could even move down to a niche market because of price competition. Pocket PCs are in the $300 to $600 range; Palms are in the $150 to $500 range. There's a lot of overlap in the price, and the Pocket PC is a much more capable device. That's a good and a bad thing; it's a much more complicated device, and if you're looking for a calendaring, simple e-mail, expense report machine the Palm may be it, because it's simple. The Pocket PC has capability to do a bunch more things. Fujitsu has been the dominant player in the tablet market for a long time; that's about to change, I believe. Maybe not their dominance, but the fact that there was no one else. Many vendors are threatening to release a machine; ViewSonic has already released one, and several other players have machines, and ViewSonic's machine is significantly cheaper. So I expect the competition in that area to change pretty rapidly, and we'll see some price reductions in those areas.

Convergence of technology seems to be on its way; the capabilities of these machines are all being put together. There are things that you can do on all of them. The Buck Rogers idea might be that your cell phone will be more capable than these other devices in a few years. That's likely to come, and the capability to run a large operating system like Windows on very small devices seems to be coming. I think that if battery consumption and speed weren't such big issues that it could fit on these smaller devices. The data storage is a big thing; you need probably a gigabyte of storage to run one of these big operating systems, but that's not far away from these devices. You can buy big plug-in non-volatile memory for these machines, so it's something that you can look at or guess that this capability will continue to grow, and it will be more capable within the same machine. It may be that you can even expect a Windows environment or a Linux environment to be available, that you can run on these devices; I think I've recently seen a Linux-based PDA on the market.

Finally, this is sort of a price review on the machines we brought with us. The Tablet PCs are pretty expensive. And these are ballpark prices; these are not GSA prices or what you could negotiate if you were buying a large number of them. But it's suggestive. Some of these handhelds—these purchase prices were five or six months old. This Casio is an industrial model of a Pocket PC; it's got a rubber case, you can drop it, it's got plugs so it can be used in the rain and won't get wet. But it's *very* expensive. These Palm PCs are down in the $150 to $350 range. As a sort of checkpoint, the Apple MessagePad was $1,000 to $1,200 when it came out—and it's not as capable as any of these devices except that, as we're seeming, screen real estate is pretty important, and this has a nice-sized screen. It's only black-and-white, so it's not as visible and you can't code things in color. But that's a basic review of the handheld price market and the tablet market at this point. I guess we could put the ViewSonic out here, at $2,000, and that's just—we saw that vendor a month ago. Maybe not even a month ago, a few weeks ago. So I think there will be pressure on these prices.

Questions?

PARTICIPANT: Have you done any usability testing with your interviewers? I'm just curious; is there any point at which small is too small?

LEVINSOHN: We have scheduled focus groups in about 10 days, and we have two concerns: one, the question is, will you go in the field with this device or this one? And we're asking: can you use it? Will you be comfortable doing that? Do you want to carry a big device? We're going to ask whether they would do doorway screening with this; is it too heavy, would you use it. So we're going to ask our interview staff. There is also concern that these screens are not as good or as big as a laptop—you can [get] a laptop with a 15- or 16-inch screen now. We're in the field now with a Gateway laptop with a 14-inch screen. And, at the time, that was a big, capable screen. These are 10-inch screens, so they're smaller; so they're going to old age homes, to collect elderly people together, have them do the interview on this and see if this is OK. Or how does this compare with doing the interview on the Gateway laptop with the bigger, brighter screen. I think there are two ergonomic issues that we know we are concerned about; the focus groups may tell us that there are some more. There's certainly the durability issue, too, that we haven't gotten a feel for. Yes?

PARTICIPANT: [*inaudible; essentially, speculating that it seems like the best option now would appear to be the tablet PC, with the hopes that the units would become more cost-effective in time*]

LEVINSOHN: I think that's depending on your time frame; for this survey, we need a new device on January 1, 2004. So I'm not sure I'd be willing to do that, because we'd need to buy that device six months—depending on how bold you are, you'd back it up four to twelve months. In that time frame, I don't know if I'd do that; in longer time frames, I really would. And porting—it depends. There is no Blaise for a Pocket PC; there, you could certainly write your own questionnaire administration tool. The tools, I believe, are strong enough to do that, but that's a big development activity. So, I think that you might decide that, well, I'm going to develop for Windows and, if worse comes to worse, I'll be working with something that's bigger than I want it to be. You could take that path. For a lot of doorway screening, you might have to convince your interviewers to carry that extra weight. But, we're carrying both devices now.

PARTICIPANT: [*Unintelligible*]

LEVINSOHN: The health study is a self-administered questionnaire, so that these devices would be turned around and given to the respondent to enter data on. I think these devices, certainly, would work with some training. I don't think using the pen for entering a lot of information with a first time user is a great idea; there's some training involved. It's sort of like using a mouse if you've never done one, or using this little pointing stick in the middle of an IBM keyboard. First half-hour you use it, it goes all over the block … So it depends, some. If you're going to approach somebody for the first time and hand them one of these devices and say, "Please answer these questions. It's very sensitive information, so I don't want to see you do it," they may need some training on how to use the pen. Or you may want to develop your interface where it's very simple and they just touch the odd thing. But I think that these devices are fine for collecting small amounts of information where, at this time, you don't need a questionnaire development tool like CASES or Blaise or something else. There is this EntryWare package; there are probably a couple of others you could do devices on.

You could put bar code scanners on this. So, these are beginning to show up in industry for sort of vertical applications, inventory applications. There, they do wireless transmission back to the database. So I think that there are … we'll see tools for that. Yes?

PARTICIPANT: For PDA-class devices [*unintelligible*]

LEVINSOHN: Come again?

PARTICIPANT: For PDA-class devices, how important do you think color is?

LEVINSOHN: I think it's important; it gives you another dimension. You don't have much screen real estate, and I think one of the changes between designing for these 14-inch screens, 10-inch screens, and the 3-inch screens is that you give up things that you were used to doing. Even the width of your slider bar becomes an issue. Color gives you another encoding dimension, so you can use color to say, "red things mean this, blue things mean this." And I wouldn't like to give that up. I think it also improves visibility. The Compaq screens and the new HP handheld screens are very sharp and crisp with their color displays.

GROVES: I have sort of another question, I guess. You've had the Newtons for nine years …

LEVINSOHN: Five … they've been out for nine …

GROVES: Oh, OK. But, do you have a sense at this point in time that you're getting better at predicting what ought to be on the next platform—what the winner is, what the loser is, and how far behind the cutting edge we should be to be wise in this?

LEVINSOHN: Well, I think it may be a little easier now than it was five years … there wasn't much else. So, if you wanted a handheld five years ago, you were going to pick a Newton. If you want one now, I think you [should] say: do I want a Pocket PC? Do I want a Palm OS? Do I want a Windows-class machine? So I wouldn't pick a vendor; I wouldn't recommend that. But I think you can pick a Pocket PC standard, program to that, and then choose from whatever vendors have survived. HP bought Compaq or vice versa—one of those two devices may disappear next year. And they're both probably—in my mind, at this point—at the top of the stack in handheld devices. But I can't imagine that they're going to continue on producing two and compete against each other.

But moving the software is pretty easy; anything we've developed so far, in our prototypes, moves from any one of these devices to another with a recompile, at worst. Sometimes you can just move the code if they're the same chip. So I wouldn't pick Cassiopeia, or Compaq—I'd say we're developing for Pocket PC or that we're developing for Windows and hope to get small enough—we're going to rely on the ingenuity of the industry to give us a device. I think you need to … if you're making that guess, you may want thirty months for that kind of device to emerge. The Pocket PCs right now are capable of doing what we're doing with the Newton, and more, and thirty months from now they'll be much stronger.

I don't think that that perform[ance] factor will go away or—if it does—it will be back up to the Windows environment and you can port your software backwards with more opportunity. I think it . . . I've been uncomfortable since day one with the Newton because of the fact that there was no manufacturer and all the software developers were finding something else to do. Apple has continued to maintain the hardware, which has helped us. And we do, we send four or five a month back to be repaired. And without that things would be getting even more nervous now.

So I wouldn't recommend saying, "Well, let's buy a lot of them and they'll be good for five years." We've done that, but I wouldn't like to do it again.

CORK: If we could hold off on other questions for right now, let's thank Jay. We'll take about a fifteen minute break while we set up for the last discussion. Jay and Marty willing, we can move the handhelds over to that table for people to touch and feel, and as long as we don't confuse the Apple Newton with the Chicken Tuscan sandwich we'll be fine.

PANEL DISCUSSION: HOW CAN COMPUTER SCIENCE AND SURVEY METHODOLOGY BEST INTERACT IN THE FUTURE?

CORK: To put a capstone on this workshop, what we wanted to do was to assemble a variety of different opinions and different perspectives to reflect on what was discussed at the workshop, and to provide their own views on the themes that were raised. And, then, to open the floor to some more general discussion. The moderator for this panel—I'll let him handle the sub-introductions from there—is Mick Couper from the University of Michigan.

COUPER: Thanks. Well, I guess you can hear me—I hope you can hear everybody on the panel as we talk. I'm simply going to follow instructions, which I'm usually not good at but I'm going to do my best at doing so. And what we have is this. Each of the panelists will have just a few minutes to give their particular perspective—about five to seven minutes—give their perspective on material both of yesterday and today in the broad theme of, "Can we all get along?" The broad theme is: what can we as survey researchers—and all of the panelists here are really survey researchers—learn from computer science in the broad sense, particularly the material we've heard in the last day or two. We will probably range all over the map, and once we've spoken we'll open things up for more discussion ... "stump the chump," or take comments or reactions from you—however you want to do it.

There's probably no logical order to go through this so I'm going to do it alphabetically. And what I'm going to do is briefly introduce the panelists so you know who's up here, then we'll go in alphabetical order with remarks. So, first, if my alphabetical system works ... to my right, Reg Baker from MS Interactive ... That's the B's. Bill Kalsbeek from the University of North Carolina. Tony Manners from the Office for National Statistics in the U.K. And Susan Schechter from the Office of Management and Budget. So we will first hear from Reg Baker.

BAKER: Should I sit here?

COUPER: Sure ... or if you want to stand ...

BAKER: Do people have a preference as to whether we stand ... does anyone care? [*Audience rumblings*] Stand on a chair? Everyone but Bill?

KALSBEEK: Tough group ... [*laughter*]

BAKER: [*moves to podium*] Let's see ... Someone yesterday characterized this session as one group of people saying it's rocket science and another group of people saying that it's not. Which pretty much, sort of, I think typifies what happens when you get anybody together with computer science people, who are very solution-oriented people and who always enjoy looking at problems and saying, "Sure, we can solve that. It's no big deal. It's really very simple." Whether or not that's the case, we'll talk about in a moment.

But I think, in my own case—because my self-image is that I'm kind of part survey geek and part gear-head—is that I just instinctively believe that sessions like this—people getting together and sharing perspectives—are a really good thing to do. And I hope that—you know, addressing the specific thing that the panel is supposed to talk about—that we can move down a road where maybe we make more progress in the future than we've been able to make in the past, in doing a better job of borrowing from what we might think of as the gear-head culture, the comp sci community. However you choose to describe it.

But first I think there are some issues that the survey folks need to address. Let's call them readiness issues, if you will. Bob Groves talked the other day about having been in a number of sessions like this; I've been in a fair share myself. And, increasingly, I sort of am reminded of the old joke, "How many psychiatrists does it take to change a light bulb?" And the answer is, "Just one, but the light bulb has to *really* want to change." So I think that, in this case, the survey group is the light bulb. And, you know, if you're going to take this seriously, there's a lot of change that's going to have to go on. And, in particular, a couple of areas occur to me.

In the opening remarks that Pat Doyle had the other day, she talked about how we used to have these "loosey-goosey" questions. And now, with technology, what we're trying to do is we are trying add more precision. But I think there are all these symptoms that we still have a pretty loosey-goosey process that's out there, trying to produce all this precision. And that the first order of business is really to go to work on that process and to try and clean it up. And that, a lot of what I hear, in terms of the sorts of the problems we have, are really problems of management and of management discipline—when I hear there are tools the people can use and that they choose not to use them. Or, we're pretty good about doing something until there's a lot of pressure and a deadline; then the wheels come off, and then everyone stops doing what they're supposed to be doing. So there are some pretty straightforward management challenges in there that I think people know what to do.

But, more importantly, I think—we tend to get together and look at our part of the elephant. And there's this whole business of a survey, particularly in the federal statistical establishment, that's much larger than that little piece that we see. And, in particular, I think stepping back and realizing that not *all* of the problems we're talking about here originate in Suitland ... some of those problems originate in downtown D.C., some of those problems originate in Hyattsville. And that there's a need to somehow get all of these people into the process, and to create a perspective for people working in it about what it is that we're really trying to accomplish here. Because, at the end of the day, what we're trying to do is we're trying to produce usable, good quality, reliable data. And I think that's easy to get lost when people are down in the bowels of these large surveys. So, I guess what used to be the appropriate metaphor in this town is, "It's the data, stupid." So when you look at decisions that you're trying to make, you look at them in terms of: is this particular change going to contribute to that goal of producing reliable and usable data?

The second thing is, I think, in all this is to reduce complexity. I mean, I was very amused—as was everyone yesterday—at those McCabe graphs which showed the complexity and really drove home the point that complexity almost invariably leads to instability in systems. And so we need to think about reducing complexity. A good friend of mine who's a consultant used to say that the thing about survey researchers I've noticed in working with them—he was not a survey researcher by trade—is that they seem to derive great pride from managing complexity when they should be deriving pride from *eliminating* complexity, from simplification. A case in point here, I think, is looking at the whole "his"/"her", "was"/"were", "a"/"an" kind of problem. And again, "It's the data, stupid." What is that really adding? And how much time and energy should go into that kind of thing?

And I was reminded—which is a little off-track—but about ten years ago I remember talking to a group of folks at a research conference about the sorts of decisions we needed to make about what sorts of systems we were going to develop for interviewers. And the difference between informating systems and automating systems, and that those decisions really get down to: what is your philosophy of your workforce, and how are you trying to get them to use technology? Automating systems are sort of systems for the factory floor, blue-collar technologies—they're all about controlling people's behavior. Informating systems are more designed for professionals—really have more to do with empowering people, using technology to give people information so that they can do a better job at whatever it is we've chosen to do. It seems pretty clear that these systems have gone down the route of automating—of turning face-to-face interviewers into what we've done to CATI interviewers, which is the automaton route. And I wonder about the wisdom of that—it may be too late to turn back. But—and, by the way, as *everyone* knows, I'm certainly a person who's in favor of eliminating interviewers altogether—but as long as we're going to have them, let's take advantage of them, and the richness of what's traditionally been the relationship between interviewer and respondent. And let's not build systems that get in the way and reduce people to simply reading things off the screen in an automatic kind of way.

As for the gearheads … what I urge on that front, for starters, is how about a little humility? I mean, let's face it—it's not an industry which is known for continually producing error-free, on-time, on-budget product. And, so, the degree to which we can sit and proselytize and help people understand that adopting our methods will solve all their problems can be a bit of a stretch. But that's not to say that there's not a lot of value there.

So, I think, really, that there are two points to be made. Number one is to be good consultants, to take the time and learn the business. I really wonder, as I sat through the discussion yesterday, whether this analogue of the interview and systems development is really as close as we like to think they are, and therefore what does that say about the applicability of the tools in the two different environments. I liked Mark Pierzchala's talk last night, which—as he said—was to bring us down to earth. And it did a good job of that, I thought. But, in general, I think that this problem of people working in a software system like Blaise or CASES, and the degree to which that's the same as doing complex systems development in C—with developers and professional programmers rather than with people who may *not* be professional programmers, and with specification systems and everything else—that we have to look really closely at the degree to which those two problems track and have the same kinds of solutions.

And, then secondly, I think, we have to be careful not to portray methodologies as silver bullets. We are very good at—"we" being in this case the gearhead side of me—as somebody pointed out yesterday, at evolving methodologies once the current methodology becomes unsupportable. And so, yesterday, we got the latest panacea: extreme programming, and all the problems that it solves. And it sits out there—when you look at the face of it—and says, gee, we've got all this chaos but *we don't need to solve the chaos*. We can work within the chaos and we can still produce good stuff. And I think that's not really the point of extreme programming, any more than it was the point of rapid prototyping. It is, granted, it's a way to come to grips with the fact that there's pressure to produce enormous amounts of software, with functionality that mostly no one wants or needs, as rapidly as possible. [*laughter*] So you have people like, you know—not to pick on Microsoft, but it's *amazing* that someone from Microsoft has been here and there's not been a single snide remark about "Evil Empire" . . . [*laughter*]

COUPER: Yet . . .

BAKER: You know, I think Mick's probably going to solve that problem when he gets up here. [*laughter*] But, nonetheless, I think you have to be really careful about recommending that people look at techniques like this because they take a lot of things that, *well*, us folks on the survey side don't have in terms of infrastructure, in terms of skill, in terms of money to be able to invest in the kinds of systems we're talking about. And, most importantly, in terms of the kind of culture that you have to create in order for methodologies like that to work effectively, not to mention how long it takes to get to where they do what it is that they need to do.

So, overall, I think that this has been really a fascinating conversation. Obviously, I thought most about it last night and this morning. But then, sitting here this morning and listening to three people talk about *new* technologies, it just occurred to me that everybody's probably getting very excited . . . Well, people can't seem to get too excited over Web; I don't know why that is . . . [*laughter*] But I'm sure that everybody loved—I know that everybody loves those little handheld devices; they're very cool, everybody likes that stuff. What I *understood* about the—what to me will forever be—the "Never Lost" system for interviewers was also very interesting. But that's been kind of the extent of the transfer we've seen as far as technologies are concerned—the devices, the capabilities, but very little on the side of how, then, you have to change your processes and how you have to manage development to really take advantage of what these technologies have to offer. That's what this is and ought to be about, and it's a direction in which I hope we can continue to go. So, thanks very much.

COUPER: Why don't we hold any questions or comments for Reg. That means you have the burden of remembering what Reg said when you want to come up with a comment. But I want to be sure to give everybody a chance to speak first before we do this. And you will learn as we go along that we're going to be all over the map. So, this will give you a flavor of the variety of issues that we covered. So, next up, is Bill Kalsbeek from the University of North Carolina. And Bill has actually got visual materials ... contrary to instruction.

KALSBEEK: I'm going to take my prerogative and use my time to introduce an angle, if you will, that hasn't really been a focus in this conference. There's been a lot of discussion in the general design of surveys and talking about the utility of the automation process—of finding people, collecting data from people, moving data, analyzing data and so forth. Not a lot has been said, however, on the very front end of surveys. And I talk specifically about the process of constructing the list of households for sampling at the final stage. In area sampling, which is used in most major household surveys in the U.S., once one gets down to the final stage, the traditional approach has been in a lot of these surveys to identify a relatively local area—below a block group, typically—and then within that area to train and send into the field field workers whose job is to basically in a very systematic way go through the area that's so designated and construct a list or a sampling frame of housing units for purposes of sampling and selecting the final sample of households. I'm from North Carolina and, in our state, what we're attempting to do is to mount an effort to produce an annual ongoing longitudinal survey of households in our state. And, as usual in these economic hard times, we're looking for ways to do this that are budget-friendly. And so there's been a lot of thought in the last 18 months as to how we might do this in a way that makes maximum use of resources. And it's covered all phases of doing a study. As everybody knows, doing a face-to-face survey is very expensive. A significant part of the budget for the field work in a household survey that uses this sort of manual field listing is that listing operation. So we began to think about ways that we might be able to save on that feature, that facet of the operation. And what we ended up doing is sort of an extension to what Sarah was talking about earlier, in the use of GIS technology. I have a very low-tech overhead that I wanted to share with you—it's actually taken from a PowerPoint presentation.

I went through and began having some very fascinating conversations with a GIS shop on our campus at the Carolina Population Center, and my colleagues there—Steve McGregor and Steve Walsh—have introduced me to some very intriguing possibilities. And the possibilities kind of built from the general notion which I have depicted here on this slide, actually a slide from one of Steve McGregor's presentations. It depicts how the layering, and the interface of all this layering of different two-dimensional data, as well as more characteristic-type statistical data can be joined. And with all the different possibilities—some of which we heard in Sarah's talk—we began to think of ways in which we might be able to accomplish this household field listing task utilizing GIS technology. An idea, as it turned out, basically was this: if we are able to use data from the Census TIGER files, and overlay that to information that is available through county property tax law offices, using property tax parcels, that the possibility might exist for sort of circumventing the need [to identify and canvass] an area (say a block group). [Instead, we could] construct a frame by interfacing or overlaying the property tax parcel—which is now rapidly being converted into a GIS format in local county property tax offices—and to utilize that interface to construct the list.

And, so, in our talks with the GIS people we developed a little plan. And what we ended up trying out—and I don't have time to go through the results of the field test that we did—I'll call it a modest field test of this idea. What we in fact did—on this very busy-looking map, a little bit too small for you to see in back—in essence what we have here is a green depiction of a block group and overlayed to that are actually the property tax parcels for Orange County, North Carolina. This is for a community, Carrboro, which is adjacent to Chapel Hill. Now the thing I wanted you to see in all of this is, first, the overlay of the property tax parcels to the TIGER maps, and then to also note the link to this of some data that was available from the property tax offices that provided information, in essence, on what they had—things like number of buildings on the property, what the property value was, the type of structure on the property, and so forth. Information that the property tax office on the county level might have. The idea here was, utilizing the GIS software that was available to us, we found that we could construct a frame, and use the block group—which as I was saying, one of the block groups on this map is basically *there*. We were able to point-and-click to construct a listing of parcels—not of individual housing units but of parcels. So the question then was … a number of questions emerged. Could this serve—the list of parcels, that is, for a selected area—could that serve as a plausible frame, or the basis for a frame, for household selection?

So what we ended up doing was developing a field experiment which we conducted in one of the counties adjacent to Chapel Hill looking at a number of questions that we thought might be questions that needed to be solved before this idea could be implemented. Questions such as: once you select a sample of parcels, are you able to send someone and are they able to identify—are they able to readily identify—the housing units that link to the parcel? This linkage, in frame construction and sampling, is the essence of the enterprise and if it can't be achieved this won't work. So one of the things we wanted to find out was whether it worked, and what we found basically is that it *did* work. We constructed an experiment based on a design where we did each task to be tested independently, twice, and then we did a comparison, and we looked at the agreement, if you will, among the tasks and the comparison adjudicated the results to see how effective it was. We were able to locate them.

Another question that came into mind was: if you construct a frame like this, you're going to have a number of these parcels which have nonresidential structures, parcels that are vacant, parcels that have businesses on them—are you able to, in any sense, delineate or isolate those that are more likely to have residential structures present? And again the answer was—and there's some uncertainty in this due to the variability in the amount of information that the property tax offices collect—the answer is a qualified yes. And the reason there is that the information that we were able to link into from the property tax offices includes some things such as valuation of a building. And if the value is 0, then it's a fairly safe assumption that there's no building there. Or, in some instances, the number of buildings is actually recorded, and so you can identify a vacant lot. In some instances, and in the case of Orange County this was true, they provided a descriptor or classification of the parcel to identify whether it was residential or a public area or the like. So you can do some limited screening to narrow things down. So, again, to the question of whether you would be able to sort of isolate more residential-type parcels, the answer was a qualified yes.

There were a number of other questions, but just so I don't completely run out of my time allotment ... The other essential question we thought we needed to answer was: given that you have a sample of these parcels, and you send people to the field, will they be able to locate the correct person? That was probably the key question for us to answer and, again, our data from this limited field test suggests that the answer is yes. And how did they do that? Well, somebody said it earlier—in rural areas, we found that the primary utility was in the GPS device that we provided. In addition to the information on the property tax, these GIS files from the property tax provide things like the centroid of the parcel, so you can use a GPS to actually [place] it. And it was remarkably accurate in the instance where we tried it in the field. So, in the rural areas, the GPS was very useful. And in urban areas we found that maps are more useful for that exercise.

So, that's about all the time that I have, but I wanted to ... I do have some things I want to say about the speakers as well, but I wanted to throw out into the discussion this other possibility dealing with the front end of the survey as a potential that could be utilized by technology.

COUPER: Thank you, Bill. As I warned at the beginning, this panel is really going to be all over the map, *literally*. Next, we're going to get a perspective from the U.K.; Tony Manners is going to add his comments.

MANNERS: Yes, literally, I'm from another part of the map altogether. And my brief here is to try to give a picture from another country, where the assumptions we work under are a bit different. Just to give you an idea about ONS, the Office of National Statistics in the U.K., what scale it's on so you can get a picture: we have 1,200 interviewers. We do, I think, 600,000 household interviews a year. One-in-a-quarter million adults interviewed, something of that order. We have about 20 projects during the year, of which about five are continuous surveys and the others are ad hoc surveys. And the variation that we have to deal with is ... our Labour Force Survey, which is like your Current Population Survey, is a 40-minute interview. The clients can get, can change 10 percent of the content every quarter. A more stable one is the General Household Survey; that changes about 30 percent of the content each year. Our Omnibus Survey changes something like 80 percent of the content every month. So, with that kind of range, we've always been looking for automated testing. Well, I should say we've been in the CAI field for more than a decade; we've always been Blaise users.

Looking for automated testing systems ... we've never really found anything that didn't take more effort to set up than the risk justified. So we've tried to concentrate on the other end; we've tried to squeeze places where errors occur, concentrate on that end. And that relates to some of what we heard yesterday. I'm going to talk about integrating survey processes, re-using code, standardizing, and—as Reg said earlier—"keep it simple." Those are the areas that we've concentrated on.

The thing about integrating survey processes—I won't talk at length about this because I've talked about this here, before. We don't have a problem of miscommunication between researchers and programmers because the researchers do the programming. Our researchers are people who negotiate with the clients, write the instruments, organize or manage the project during the fieldwork, and analyze and write the reports. So, they've got an overall picture.

We do a lot of standardization. One of the things that strikes me in the U.S. is that you seem to build case management systems around particular projects. I mean, we just have *a* case management system that projects slot into. Maybe I've misunderstood ... We do a lot of standardization of code. We have a big initiative in government—across all departments in government—to standardize all of the basic classificatory information. And, because Blaise is such a modular language, you can easily produce modules which people—these researchers—can literally assemble into questionnaires.

One of the things, actually, that has been remarked on is that we talk about complexity all the time. Actually, our software has gotten better in the sense that it enables you do things more simply; you just have to sure that your aim is guided simply. And that's been one of our uses of the better software is that we don't do more; we actually try to do less.

Something else that I think is worth mentioning, because I haven't heard much about it, is how important it is to look at the whole process and to build the output structures into your instrument. You build things up front, but you've got to be able to deliver data to people. And, in our experience, one of the areas where that potentially goes most wrong is between the output of the field instrument into whatever databases your clients are using. So we devote a lot of effort to persuading our clients of the virtues of simplicity. We had, on our Expenditure and Food Survey—which is a little like your Consumer Expenditure Survey, but it's got nutrition as well—that we produced basically 99 tables out of the instrument. It was literally 99, and it was a painful process. We persuaded them to move to 3, as the natural structure of the instrument.

Something else we do . . . we write deliberately inefficient code. Code, that is, which is inefficient in programming terms but is very efficient in terms of the overall process. We do things so that the whole organization can understand the code that's been written. For those who use Blaise, for example, we don't use parameters—because that's too much like real programming.

Like I said, we re-use code. We have a list of standard modules, and we also have—in our ad hoc surveys, something like 80 percent of the survey will be completely new. But the structures aren't new. So what we do is have people use templates, which use the structures they want to use, and again assemble those.

Standards . . . our interviewers work across a range of surveys, as is the case with many of yours, and obviously they've got to be looking at the same kinds of screens all the time. So you have to have standards for things like that. But we also have standards for writing code and so on. We have a small group—one full-time equivalent, but it's a number of people—called the Standards and Quality Assurance group, which kind of keeps a check on the fact that people are actually following the standards.

I've spoken about "keeping it simple." We do that literally, in terms of—if people want to anything *different* from the standards, from the norm, they've got to make a business case for it. You have to be very clear that something you add adds value.

Coming back to the things that we heard yesterday ... yes, I think, Pat Doyle talked about enforcing specifications. Or getting agreement on specifications. We spend a lot of time on that, and trying to draw clients in so that the client understands the benefit that they will get from very clear specifications. So, it sounds obvious but it's not always what it seems to be.

Something Bob Groves said ... I'd just like to note that my organization has made very significant savings from moving to CAI. I think that that's not normally the case, but ... the reason we've done that is that that was our initial motivation. We had to save money, and so we designed them to save money.

Thomas McCabe—I very much like his index of unstructuredness, and I plan to use that.

With Harry Robinson, I agree very much with his idea about trapping as many bugs as the time and resources allow. That sounds like a real goal. Robert Smith, I agreed very much with. And a lot of what he said about prototyping, design restraint, modularity, transparency, and building quality in from the start. Larry Markosian, the same, in terms of iterative development and involvement of the stakeholders throughout—I think that's a very important point. And with Mark, of course, I agree about coming back to Earth, and I liked how he drew attention to the sort of "fuzzy" things that we have to test.

So, I haven't said a great deal about our testing process but I suspect it's much like everyone else's, with an awful lot of manual testing and people trying to do that in as intelligent and structured way as possible. But, as I say, we concentrated more on the places where errors might occur. So ...

COUPER: Thanks, Tony. Next, we'll hear from Susan Schechter from the Office of Management and Budget.

SCHECHTER: I'm going to spend just a few minutes talking about the OMB process, from an official perspective, and then I'm going to give you a couple of my opinions on how that integrates with survey automation. And then I'm going to mention briefly two initiatives that I think will continue to affect the submission of collections to OMB, from the agency perspective.

Most of you know, but I'm not sure that all of you know, that the OMB process basically requires every federal agency that wants to collect any amount of information from one or more persons in the public— whether a business or a school or a household respondent—must come to OMB to get permission to do so. And that's called the clearance process. The main objective of the agency is to get an OMB clearance number and an expiration date. And the expiration date *typically* is three years, which means that once you get OMB clearance approval you don't have to come back to OMB for three years.

This process, for agencies, is agonizing. And I used to be in an agency, and I've worked that process, and I know how agonizing it is. Listening to Pat's talk yesterday and Jesse Poore's talk, probably to me the biggest problem is that OMB comes in right at the end of the process. And they have not been involved in the conceptual design of surveys, in the development of the questionnaire—typically. There are exceptions, but typically. Yet it's OMB's responsibility to review and approve the survey.

I want to mention to you—I don't know how many of you know—of all the data collections that OMB approves during the year less than 5 percent are sample surveys. The vast majority are reporting forms. In fact, 80 percent of the entire burden that OMB approves during the year are IRS forms. So, we're not talking just about a process that fits just the survey world; we're really talking about a process that fits to a tax report form, a birth form, any form of administrative reporting, a federal loan application, etc. And all of this process has to fit within these different types of forms.

What ends up happening when the agency wants to finish its development of work—the agency does have to tell the public in a *Federal Register* notice that they're getting ready to send something to OMB. And the public gets 60 days to comment on the agency's notification of a proposed information collection. Then, after that 60 days closes, the agency comes again to OMB, this time with a *Federal Register* notice that says, "now, we really mean it and we really want to collect some information." And you have 60 days to tell us if you have any objections or concerns about that. And after that 60 days, that's when OMB is supposed to act.

In the past, paper-and-pencil forms were most common and in many cases they are still most common. But, we will talk about surveys for a second ... OMB had a fair amount of discretion during those 60 days because the form is still sitting at a form design desk without being printed, because it didn't have a clearance number and an expiration date. When surveys transitioned to automation, that's where it changed. And it changed dramatically, and I'm not sure OMB still realizes just how much it has changed. Because the truth is that, by the time a survey comes to OMB, there is very little flexibility for OMB to change something in the instrument. I would say that in most cases agencies are tremendously resistant to any change in wording, in response categories, in question order, in something very minor where OMB—a policy official in OMB—may go, "why don't you ask this question?" One question, on something like the CPS or the SIPP, would cause probably a year in advance that the agency would need notification in order to ask that question. Because there's this testing process that Pat illustrated; there's the documentation process. There are all kinds of issues about field manuals and programming, and all the pieces of the puzzle that go into building a survey. So I would say that OMB does the best it can in its function to review instruments and data collection, but I would say that the focus has become less and less on the quality of the survey instrument and more on the utility of the survey, the design, the methodology. Yes?

MARKOSIAN: I didn't understand, perhaps I missed at the beginning ... what are the criteria that are used and why is your agency involved in this process?

SCHECHTER: Ah, yes ... The Paperwork Reduction Act specifies that agencies should really try to stop burdening the public with all of this collection of information. And so if you have to burden the public, you have to tell OMB why you need this information. You have to demonstrate the practical utility of the information, you have to demonstrate that you've reduced the burden as much as possible. It's sort of a way to manage what is perceived as this huge burden to the public. And in fact, a report just came out that said that it's something like 7 billion hours a year that the public suffers from. But much of that, 80 percent of it, is IRS forms. So that's why ... [*laughter*] So it's a way to try to reduce ... agencies wholesale asking people for information.

So when Pat showed you yesterday, for example, the variation on the children's health insurance program question, the truth is that OMB rarely sees all those different versions of the questions. They think they're seeing the question that's going to be asked—you don't always see the question in all the different modes, you don't always even understand all the different modes that are being asked. The agencies certainly have a responsibility to document their methods to OMB but a lot of that depends on the familiarity that the desk officer has with survey research. Some of the desk officers work much more with forms, not with surveys, with administrative kinds of data collection, and they are trained in policy areas, in economics, etc. They are not survey methodologists, per se, although some are. But most are not. So there's a tremendous amount of variability in the desk officer role.

I would not say that OMB should move in a direction where we insist on more and more documentation from agencies. I wouldn't like to really see that happen. On the other hand, I'm not sure what our responsibility is when we're told that the survey may be completed on the Internet, for example. And that the agency may offer that as an option. Should the agency be sending us all the screens that the person is going to see? Should the agency be sending us a test demo, or a link, where we would see that? There are some desk officers that ask for that, but I would not say that that is a routine kind of request that we ask for.

So my big, I guess, conclusion is that in terms of automation we are not consistently reviewing all of these submissions in the same way. There are some submissions that are reviewed much more carefully. The documentation that we see usually has—if it's a CAPI or a CATI—usually has a coding that shows some skip patterns or some GOTO instructions, sometimes not. Sometimes you just get lists of questions. And we have to review the list of questions and assess whether that's enough to approve it, or if we have to back to the agency and say, "could you give us more to help us understand the real content of this survey?"

I want to tell you about two initiatives that I do think will affect the agencies in coming years, one already passed. There's an initiative called GPEA, which you may have heard of, it's the Government Paperwork Elimination Act. And the essence of this act is that every data collection must offer the option to a respondent that they can reply in an automated fashion. It doesn't mean you *must* get rid of paper-and-pencil surveys; it means that you must offer an option to report either on CD-ROM or on Internet or in some kind of electronic mode. Not all surveys have come into line with that, and I don't think we have a lot of pressure yet to force it, although we are certainly expected to be asking the question: if this is a manual survey, when are you moving to an automated option? I do think that in the future that's going to become a bigger issue. And I think that there will the difficulty for some surveys to do that. For example, there is just a resource constraint. I'm not sure, for example, what the current status of the National Crime Victimization Survey is, but the last time we reviewed it it was still paper-and-pencil because they didn't have the money to pay for the conversion to CAPI. So if OMB is not going to provide that money, and yet there's a legislative requirement to offer an automated reporting option, I guess at some point there's going to have to be a meeting of those minds.

I've been told that a CAPI interview will qualify as an automated reporting from a respondent. But my guess is that agencies will have to demonstrate that an interview is so complicated and so complex that you have to have an interviewer's help. They can't just tell them to do it on a telephone data entry or you just can't give them the Internet option. I don't know how that is all going to play out.

The other interesting thing that's happened this year—and I'm not sure that we're going to really see the effects of it for awhile—is that there's been a Data Quality Initiative that we put onto the OMB appropriation language last year, which ended up requiring OMB to put out data quality guidance and requiring that agencies put out data quality guidance. And I have to step back from that from a minute, and I will come back to it. When OMB approves a survey, OMB never sees the survey again. We approve a survey and then the agency says that this is the questionnaire we'r going to use, this is the methodology that we're going to use, and this is the response rate we think we're going to get. We don't say, "now, come back and tell me next year how you did. And give me a copy of the report showing the data." Sometimes we do, if we have personal interest, but generally we do not. And unless the agency substantially changes the content or the sample design—-something *really* substantially changes—they don't have to come back to us. So we only approve up front. The Data Quality Initiative only cares about the end—the end analytic result—and whether or not the findings are based on high quality data. Now, of course, it does really care about the process and the development of surveys and the data collection. But I think what's going to happen is that agencies are going to find themselves doing more and more documenting, more and more publishing what their technical notes are, what their concerns are, what their issues are about the data quality. I think that archiving will end up becoming more of an issue, because people are going to be able to go back and say, "Well, you based some rule on a study that was done that was unpublished—I'd like to see some of that data. I'd like to be able to understand how you came to the conclusions that you came to."

So I do think that, between this Government Paperwork Elimination Act and the Data Quality Initiative, it will impact the OMB process— I'm just not sure *how* it will. So I'll leave you with those thoughts for today . . .

COUPER: Thank you, Susan.

Well, I just sat there—and we're running out of time—so just to segue into the open discussion, let me make a couple of remarks. I know Harry from Microsoft is here, and some of the computer scientists, so I promise not to say anything nasty about Microsoft. Which will be difficult for me.

But just to kind of caper into the lion cage for a moment and then come back to what this discussion has been about over the past day and a half—what can we do to get together? What can we as survey researchers learn from the computer scientists? And you've heard a lot of different perspectives about things that we can do.

I would just make one assertion, just to get the discussion going, which is that the burden is on us. It's going to be hard for computer scientists out there to say, "Gee, this is a fascinating problem. I want to devote my life, devote my resources, to that." The burden needs to be on us, and I think that we do ourselves a disservice if we keep on doing this hand-wringing about how unique and how complex our problems are. And this isn't specifically at you, Pat—we all do it. If our problems are *so* unique, how are we ever going to get people in other disciplines interested enough in our problems, to devote time and energy to finding out what those problems are? So, and I think that on that point we've heard several suggestions and several comments. One of the ideas that came up, particularly yesterday, are things like the complexity graph, things like model-based testing, things like using tools that are out there, using ideas that are out there. And we can take those ideas and apply them to our subjects—in small incremental steps. I think we also fail ourselves sometimes when we think of everything as a massively complex task, and it becomes unmanageable. "We can't handle it because it's too big," and then we stop. So what I think we should do is, many of the problems and many of solutions I heard yesterday are really on the order of one smart graduate student working over the summer to demonstrate how these kinds of tools could be applied to a module of the survey— enough to get people excited about the idea.

The other final comment I would make, and then I promise I will turn it over to the discussion, is that one of the things I learned yesterday is that—despite being extremely complex, as the big surveys are—they're not as complex in the sense that we can analyze them. People showed that we can analyze them, we can draw graphs, we can generate graphs from these systems. So they might be complex but they're understand-able. The IDOC system showed that we can extract systematic informa-tion from them, to produce documentation. And if we think of survey systems, automated survey systems, in that way, they are certainly com-plex but they can be made less complex and more manageable. There are some systematic aspects of them that can be made amenable to writing scripts that will parse all the information and produce diagnostic tools. So I think we can make progress.

I'm heartened, from what I saw. A lot of ideas over the last day and a half. I think that there's a lot we can do to improve the process and I think we ought to be taking the lead, the next step, to do that.

So, with that, what I'm going to do is kind of moderate and manage the discussion. And we can slowly segue into lunch after a few minutes. So this is now open to the floor to add comments, questions, ultima-tums, or whatever you would like.

ONETO: I'll start with a question for Susan. I was intrigued about the comment you made about OMB getting into the scrutinizing of data quality. I just … I mean, you could have a survey that is just *beautifully* documented and that has very modest respondent burden, that doesn't duplicate the information that any other survey collected, and it may not have the data quality. So the question is: who is going to determine what "data quality" is? And how will OMB be judging something like that?

SCHECHTER: Right ... It's an excellent question; it's something that has us all worried. I would say that the statistical agencies themselves are probably going to come up with a common framework for how they're going to define and look at data quality. The essence of this particular statutory initiative was that rules that are passed that are very costly to implement are often based on studies that EPA may do or other agencies, that HHS may do, and then based on the results of rules they implement a legislative initiative that might require industry to spend a fair amount of money to do something. There were a lot of complaints in the last administration—remember the ergonomics, big snafu, that we got this ergonomics rule finally passed and the new administration came in and pulled it back? And part of it was that it was based on data that was not of high enough quality to justify that rule. I'm not exactly sure what the outcome is going to be for survey data. The way that the guidance is written is that if the data are *"influential"* they have a higher standard. And by "influential," it's defined as being used for policy or decision making. Well, if you think about most of the data that comes out of the Census Bureau, most of the data that comes out of the Bureau of Economic Analysis, out of BLS, etc., the data are used for decision making. There are some who think that this will end up in the courts, and that the courts will eventually get involved in some disputes about how high was the quality of the data for use in decision making. I'm not sure where it's going to go. I think, personally, that it's going to become a new area that some people will work a lot of hours in.

DOYLE: Most people might predict that the issue I'm going to raise is one of documentation ... [*laughter*] And I'm hoping that my documentation folks will back me up here. I am encouraged by the progress that we've made in documenting instruments. I am not quite as optimistic as you are that we've really solved the problem yet because neither of the two tools that we have are really ready for prime-time today. But we have needs today that need to be filled. But my bigger concern is that, with automation, because we lost the free good of the paper instrument, we need a substitute. We don't have an automated substitute, and we're never going to get an automated substitute that can provide the context for each question, why we ask each question, etc., etc. So I think we need a change in behavior amongst our survey methodologists, amongst our survey managers, amongst the managers in the organizations, about— and it applies to testing as well. It's a new behavior that we need to adopt so we can continue to provide the same information that we've always provided. So what I want to do is to hear from those of you in the education business, particularly those of you involved in educating our survey methodologists—our future managers—will you consider adapting your curriculum to address some of these issues? They're not theoretical, survey methods issues, but they're very practical management and design issues. It gets to the issues that Tom [Piazza] is talking about, about choosing a simpler approach. It may appear to be precise but it's not so precise, it should be kept simpler. But I think that the only way we can make a wholesale change in all of this is to educate our up-and-coming folks to move in that direction. So I'm wanting you guys to be enthusiastic about this and to take this task on, to expand your teachings.

PARTICIPANT: Any education people want to take that up?

PARTICIPANT: She glared at me when she said that ... I think it's not only the matter of capturing the documentation. I think one of the worst problems we've always had working with surveys, or any old data set, is that the documentation not only is incomplete but it hasn't been structured. And I think—Tony Manners said that we need to think in terms of developing surveys within a structured system that captures those decision points, that captures the conceptual points as well as the survey development. That not only captures the information but does it in a structured way so that, as information is turned in to OMB, that they know what they have to look at. Within such a structure, you could begin to define some of those complexity, those tests for complexity. You want a structured system, because that's when you can test computer programs and their structure and know what to expect. When your documentation isn't structured in a uniform way, [that's a difficulty]. So, I think it does get back to the documentation; if it isn't captured, captured completely and in a structured manner so that you can go through the process and to the data that comes out at the end and go back and say—when doing analysis down the road—how you got to this point, I think that's what needs to be worked into this whole process. And I think that documentation and testing has traditionally been something done after the fact, and it needs to be thought of [all along].

TUCKER: I want to go back to something Mick said ... is this really as complex as we make it out to be? I don't think that the survey process itself is that complex; I think that what's complex is the organization built around it. And I think we may ... it's become so complex because we're not organized to do it well. And, so I asked the question yesterday—I think it's a people problem, I don't think it's a technology problem. I think, ultimately, we'll have the technology—the question is whether we'll know how to use it.

COUPER: Let me respond to that, and I think Reg touched on it in his remarks. ... It's about processes and about people, and it touches on the documentation, about skill base and knowledge, about being able to use the tools that *are* available. So I do think that it has to do a lot with structure and management and those sorts of things. So I agree with you. Reg, you want to jump in on that?

BAKER: Well, let me just say this, and it's not going to be very popular. I think there's ... I'll just spit it out. [*laughter*] There is a tendency, I think, in the federal statistical establishment to let standards run amok. And there are lots of examples of that. You know, the one I talked about a while back at CASIC has to do with security at BLS. Well-meaning folks who managed to pretty much kill any chance the CES had to be deployed on the Web in any kind of planned fashion because of insisting that people download digital certificates. Now, who's not in favor of security and confidentiality of data? We all are. But you still have to be practical about it. I think a lot of the things you heard Pat talk about yesterday are the sorts of things ... if you have a very careful, conscientious reading of the methods literature as you're designing questionnaires it's very easy to fall into this trap of wanting to do all of these things which add a lot of complexity, which ... you know, methods studies single out. ... I was listening to Roger [Tourangeau] this morning and thought, "Damn it, this is just going to make things more complicated, studies like this." So I think there's that tendency ...

TOURANGEAU: I'll go sit outside ... [*laughter*]

BAKER: But I think that there is that tendency, and the part that's probably not popular to say is that it's because a lot of this evolves in a not-competitive environment where the need to be leaner, to be faster, simply doesn't exist. And so we can in fact spend a lot of time on standards, and insisting that standards be followed in a certain way, that maybe someone in the private sector doesn't have the luxury, and they need to look at those things from a practical perspective and make hard choices.

KALSBEEK: I'd like to comment on the issue of simplicity. I'm for simplicity; I'd like to go on record to say that ...

BAKER: Everyone always is ... [*laughter*]

KALSBEEK: ... The difficulty is, my field is public health. I work with researchers who ask difficult questions about important, difficult societal problems such as abuse in the home. So, if I'm working with that researcher and my job is to help that person to extract information from women who may be exposed to abuse of their children or abuse of their spouse, I'm touching on some very difficult ground. And it may be, and there's plenty of evidence to suggest, that if I'm that respondent I won't be fully forthright if you ask me the question in a conventional simplistic way. And I may have to find some more complex alternative such as preserving confidentiality through use of technology to remove identifying information—some states have legal implications in that regard, in terms of recording instances of abuse—and an interviewer can legally come into harm's way if there isn't some provision made for sort of developing a gap or a fail-safe so that person won't be subject to litigation. The conventional simplistic solution won't solve that problem, so I may have to build the capacity in collecting information which will complicate matters.

My point here is that to extract and to produce good data—as OMB wants us to do—in some instances, we have to use extraordinary means, and that interjects complexity, and the complexity interjects complexity with the technology, to get the job done.

PARTICIPANT: [*inaudible until microphone is obtained*] In the academic world of software engineering and computer science, there's very little emphasis on things like maintenance, support, and documentation. And everybody who has gone from teaching computer science to being VP of engineering or something—which is my route—you discover this and you kind of look back and say, "Gee, I wish I had had a course that was teaching about documentation." And I don't know how to get that back into play; it's a kind of general problem that we seem to have a mismatch on that, and we see it there, too.

PARTICIPANT: I think that the idea is to produce usable data, and "usable" is the thing that I get to, and I see it in a slightly different way from the archival community. And this relates to some things we've heard today. There are some standard ways of producing something describing what you've done; there are emerging standards, there are standards that are already in existence. I see another area, and that's in the area of cognition. And one of the steps we need to understand something that is relatively complex ... This happened when working with Pat, when the SIPP data first came out, it was a very complex undertaking with a new analytical technique. But we also had to get together and say, well, what do people need in order to figure out how to use this data? One of the things that I was told by the conference organizers that we would come up with is that in terms of Web searching and such it is necessary to sit down and think about what are the products and what do people need—not only from the organizational level and the files, but also from the cognitive level—so that we understand what is going on in the survey. [*inaudible*]

COUPER: Well, I think that we might be nearing the end ... I see you all eyeing the lunch boxes. [*laughter*] And I can understand how exciting lunch might look now. So Dan's going to tell us how to do this.

CORK: There's not much more to say ... lunch is deliberately informal today. We'd like you to stick around to chat, to converse. The formal part is all done on the workshop. I hope you enjoy the lunch. Thanks to all of you for coming, and thanks again to all of our speakers.

Workshop Information

AGENDA

Workshop on Survey Automation
The Melrose Hotel, Washington, DC

Day One: April 15, 2002

9:00 Introductions and Opening Remarks
Chester E. Bowie, *U.S. Census Bureau*

9:15 Overview: Problems in Current CAPI Implementations
9:15 CAPI Implementation of the Survey on Income and Program Participation
Pat Doyle, *U.S. Census Bureau*
10:30 Break
10:40 Idealized CAPI Implementation from Computer Science Perspective
Jesse Poore, *University of Tennessee-Knoxville*
11:20 Automation and Federal Statistical Surveys: Need for a Bridge
Robert Groves, *University of Michigan*

11:30 Lunch

12:30 Documentation of Complex CAPI Questionnaires
12:30 Understanding the "Documentation Problem" in Survey Automation
Tom Piazza, *University of California at Berkeley*
1:15 The TADEQ Project
Jelke Bethlehem, *Statistics Netherlands*
2:00 Computer Science Approaches: Visualization Tools & Software Metrics
Thomas McCabe, *McCabe Technologies*

2:45 Break

3:00 Tutorial on Software Engineering and Model-Based Testing
Harry Robinson, *Microsoft*

3:40   **Testing of Computerized Instruments**
3:40   Case Example of Software Testing
      Robert L. Smith, *(Formerly) Computer Curriculum Corp.*
4:10   An Integrated View of Survey Automation
      Larry Markosian, *Independent Consultant*
4:35   Practitioner Needs, and Reactions to Computer Science
      Approaches
      Mark Pierzchala, *Westat*

5:15   **Reactions and Floor Discussion**

Day Two: April 16, 2002

9:00   **Emerging Technologies in Survey Automation**
9:00   Web-Based Data Collection
      Roger Tourangeau, *Joint Program on Survey Methodology*
9:40   Interface of Survey Methods with Geographic Information
      Systems
      Sarah Nusser, *Iowa State University*
10:20 Prospects for Survey Collection Using Pen-Based Computers
      Martin Meyer & Jay Levinsohn, *Research Triangle Institute*

11:00 Break

**11:15 Panel Discussion: How Can Computer Science and Survey
        Methodology Best Interact in the Future?**
      Mick Couper (moderator), *University of Michigan*
      Reg Baker, *MS Interactive*
      Bill Kalsbeek, *University of North Carolina*
      Tony Manners, *Office for National Statistics, United Kingdom*
      Susan Schechter, *U.S. Office of Management and Budget*

12:15 Lunch, Final Comments, and Adjourn

LIST OF WORKSHOP PARTICIPANTS AND ATTENDEES

In the following list, the names of workshop participants—those with a planned speaking role—are preceded by •. Those invited guests and attendees who asked questions or made comments at the workshop—and for whom voices or surrounding speech made identification in the proceedings text possible—are preceded by †. Our apologies to those audience members whose contributions to the workshop could not be positively identified from the workshop tapes and whose contributions are unfortunately cloaked by the label PARTICIPANT in the text.

Tammy Anderson, U.S. Census Bureau
Karen Bagwell, U.S. Census Bureau
•**Reg Baker**, MS Interactive
†**David Banks**, Food and Drug Administration
Patrick Benton, U.S. Census Bureau
Lew Berman, National Center for Health Statistics
•**Jelke Bethlehem**, Statistics Netherlands
Chester E. Bowie, U.S. Census Bureau
Janis Lea Brown, U.S. Census Bureau
Lynda Carlson, National Science Foundation
Constance Citro, National Research Council
Cynthia Clark, U.S. Census Bureau
•**Michael Cohen**, National Research Council
Quentin Coleman, National Agricultural Statistics Service
•**Daniel Cork**, National Research Council
Joe Cortez, U.S. Census Bureau
•**Mick Couper**, University of Michigan
Kathy Creighton, U.S. Census Bureau
†**Cathryn Dippo**, Bureau of Labor Statistics
•**Pat Doyle**, U.S. Census Bureau
†**Ed Dyer**, U.S. Census Bureau
Jimmie Givens, National Center for Health Statistics
Nancy Gordon, U.S. Census Bureau
Ann Green, Social Science Statistical Laboratory, Yale University
•**Robert Groves**, University of Michigan
Doug Guan, U.S. Census Bureau
Susan Hauan, U.S. Department of Health and Human Services
Heather Holbert, U.S. Census Bureau
Bernie Greene, National Center for Education Statistics
Tim Hart, Bureau of Justice Statistics
•**William Kalsbeek**, University of North Carolina

Howard Kanarek, U.S. Census Bureau
Patsy Klaus, Bureau of Justice Statistics
Cheryl Landman, U.S. Census Bureau
•Jay Levinsohn, Research Triangle Institute
Jennifer Madans, National Center for Health Statistics
•Tony Manners, Office for National Statistics, United Kingdom
Tim Marshall, U.S. Census Bureau
•Lawrence Markosian, independent consultant
•Thomas McCabe, McCabe Technologies
•Martin Meyer, Research Triangle Institute
Bill Mockovak, Bureau of Labor Statistics
Steve Newman, Westat
•Sarah Nusser, Iowa State University
Jim O'Reilly, Westat
†Adrienne Oneto, U.S. Census Bureau
Andrea Piani, U.S. Census Bureau
•Tom Piazza, University of California at Berkeley
•Mark Pierzchala, Westat (at time of workshop) and Mathematica
     Policy Research
•Jesse Poore, University of Tennessee-Knoxville
•Daryl Pregibon, AT&T Labs–Research
Ray Ravaglia, Education Program for Gifted Youth, Stanford
     University
Callie Rennison, Bureau of Justice Statistics
•Harry Robinson, Microsoft
Johanna Rupp, U.S. Census Bureau
•Susan Schechter, Office of Management and Budget
•Michael Siri, National Research Council
•Robert L. Smith, Computer Curriculum Corporation
†Miron Straf, National Research Council
Libbie Stephenson, Institute for Social Science Research, University
     of California at Los Angeles
Anne Stratton, National Center for Health Statistics
Wendy Thomas, Minnesota Population Center, University of
     Minnesota
•Roger Tourangeau, University of Maryland
†Clyde Tucker, Bureau of Labor Statistics
David Uglow, Mathematica Policy Research
Laarni Verdolin, U.S. Census Bureau
Andrew White, National Research Council
Arnie Wilcox, National Agricultural Statistics Service

Bibliography

Baker, F.T., and Harlan D. Mills
  1973 Chief programmer teams. *Datamation* 19:58–61.
Couper, Mick P., Reginald P. Baker, Jelke Bethlehem, Cynthia Z.F. Clark, Jean Martin, William L. Nicholls II, and James M. O'Reilly, eds.
  1998 *Computer Assisted Survey Information Collection.* New York: John Wiley and Sons.
Crowne, Douglas P., and David Marlowe
  1960 A new scale of social desirability independent of psychopathology. *Journal of Consulting Psychology* 4:349–354.
Doyle, Pat
  2001 What good is a metadata system without documentation? *Of Significance . . .* 3:5–24.
Dyer, William E., Jr., and Ellen Soper
  2001 Challenges of instrument development. In *Proceedings of the 7th International Blaise Users Conference.* Washington, DC: Westat.
House, Carol
  1985 Questionnaire design with computer assisted telephone interviewing. *Journal of Official Statistics* Vol. 1, No. 2.
McCabe, Thomas J.
  1976 A complexity measure. *IEEE Transactions on Software Engineering* 2:308–320.
Nass, Clifford, Youngme Moon, and Nancy Green
  1997 Are computers gender-neutral? Gender stereotypic responses to computers. *Journal of Applied Social Psychology* 27:864–876.
National Research Council
  1993 *The Future of the Survey of Income and Program Participation.* Panel to Evaluate the Survey of Income and Program Participation, Constance F. Citro and Graham Kalton, eds. Washington, DC: National Academy Press.
  2000a *Summary of a Workshop on Information Technology Research for Federal Statistics.* Committee on Computing and Communications Research to Enable Better Use of Information Technology in Government. Computer Science and Telecommunications Board and Committee on National Statistics. Washington, DC: National Academy Press.
  2002b *Data Needs for the State Children's Health Insurance Program.* Panel for the Workshop on the State Children's Health Insurance Program, Arleen Leibowitz and Earl S. Pollack, eds. Washington, DC: National Academy Press.
Paulhus, Delroy F.
  1984 Two-component models of socially desirable responding. *Journal of Personality and Social Psychology* 46:598–609.

Prowell, Stacy J., Carmen J. Trammell, Richard C. Linger, and Jesse H. Poore
    1999 *Cleanroom Software Engineering: Technology and Process.* New York: Addison-Wesley.
Watson, Arthur H., and Thomas J. McCabe
    1996 Structured Testing: A Testing Methodology Using the Cyclomatic Complexity Metric. Special Publication 500-235. National Institute of Standards and Technology, Gaithersburg, Maryland.
Westat
    2001 Survey of Income and Program Participation Users' Guide (Supplement to the Technical Documentation). Prepared for the U.S. Census Bureau.

Biographical Sketches of Workshop Participants and Staff

Reginald Baker is chief operating officer for Market Strategies, Inc., a full service market research firm located in Southfield, Michigan, and specializing in electric and gas utilities, information technology, health care, and public policy. He is responsible for all of MSI's sampling, interviewing, data processing, and information technology operations. In addition, he is president of MSInteractive, a Market Strategies company with a special focus on Internet-based research. Prior to joining MSI in 1995, he worked for 11 years at the National Opinion Research Center (NORC) at the University of Chicago, where he was vice president for research operations. Prior to joining NORC, he served as president of Data Use and Access Laboratories in Arlington, VA. His research interests include data quality impacts of new data collection technologies, interviewer training for computer-assisted person interviewing, and future developments in computer-assisted information collection.

Jelke G. Bethlehem is senior advisor of the Department of Statistical Methods of Statistics Netherlands. He joined Statistics Netherlands in 1978 and served as head of the Statistical Informatics Department (a research unit within the Division of Research and Development) from 1987 to 1996; the department is responsible for development of software for processing survey data, including the Blaise system for computer-assisted survey data collection and the StatLine system for electronic dissemination of statistical publications. He is involved in research projects in the areas of survey methodology and coordinates European research projects. He is also part-time professor in statistical information processing at the University of Amsterdam. His research interests include treatment of nonresponse in survey data, disclosure control of published survey data, user-friendly software for statistical analysis, and graphical techniques in statistics. He has a Ph.D. in mathematical statistics from the University of Amsterdam.

Michael L. Cohen is a senior program officer for the Committee on National Statistics, currently serving as co-study director for the Panel on Research on Future Census Methods and staff to the Panel to Review the 2000 Census. He previously assisted the Panel on Estimates of Poverty for Small Geographic Areas. He also directed the Panel on Statistical Methods for Testing and Evaluating Defense Systems. Formerly, he was a mathematical statistician at the Energy Information Administration, an assistant professor in the School of Public Affairs at the University of Maryland, and a visiting lecturer at the Department of Statistics, Princeton University. His general area of research is the use of statistics in public policy, with particular interest in census undercount, model validation, and robust estimation. He has been elected a fellow of the American Statistical Association. He received a B.S. degree in mathematics from the University of Michigan and M.S. and Ph.D. degrees in statistics from Stanford University.

Daniel L. Cork is a program officer for the Committee on National Statistics and responsible staff officer for the Workshop on Survey Automation. He is currently assisting the Panel to Review the 2000 Census and serving as co-study director of the Panel on Research on Future Census Methods. His research interests include quantitative criminology (particularly space-time dynamics in homicide), Bayesian statistics, and statistics in sports. He holds a B.S. degree in statistics from George Washington University and an M.S. in statistics and a joint Ph.D. in statistics and public policy from Carnegie Mellon University.

Mick P. Couper is a senior associate research scientist in the Survey Methodology Program in the Survey Research Center at the University of Michigan, an adjunct associate professor in the Department of Sociology at the University of Michigan, and a research associate professor in the Joint Program in Survey Methodology at the University of Maryland, College Park. He previously worked at the Census Bureau as a visiting researcher from 1992 to 1994. He has published in the areas of survey and census nonresponse and the use of computer technology for data collection. He has a Ph.D. in sociology from Rhodes University in Grahamstown, South Africa.

Patricia J. Doyle is the survey improvement coordinator for the Demographic Directorate at the U.S. Census Bureau. A fellow of the American Statistical Association, she has been working with complex demographic surveys since 1976, concentrating on data base development, documentation, and microsimulation modeling. Most recently she spearheaded the Survey of Income and Program Participation (SIPP) Methods Panel Project, a 5-year project to systematically develop and test an alternative automated SIPP instrument. She co-edited the recent book, *Confidentiality, Disclosure, and Data Access: Theory and Practical Applications for Statistical Agencies*, and organized a conference on that same topic in January 2002. Her efforts to improve automated surveys include an analysis of the impact of automation on demographic surveys, design and development of automated instrument documentation procedures, and developing guidelines for documenting demographic surveys. She has an M.S. in applied mathematics from the University of Maryland, Baltimore County.

Robert Groves is director of the Survey Research Center at the University of Michigan, where he is also a member of the Survey Methodology Research Program. His research interests include survey methods, measurement of survey errors, and survey nonresponse. His current research focuses on theory-building in survey participation and models of nonresponse reduction and adjustment. He is a fellow of the American Statistical Association, an elected member of the International Statistical Institute, former president of the American Association for Public Opinion Research, and chair of the Survey Research Methods Section of the American Statistical Association. He has an A.B. degree from Dartmouth College and a Ph.D. from the University of Michigan.

Joel L. Horowitz is professor of economics at Northwestern University; he was previously professor of economics at the University of Iowa and a senior operations research analyst at the U.S. Environmental Protection Agency. His primary area of research is in theoretical and applied econometrics, with particular concentrations in semiparametric estimation, bootstrap methods, discrete choice analysis, and estimation and inference with incomplete data. He is an elected fellow of the Econometric Society, a winner of the Richard Stone Prize in Applied Econometrics, and a recipient of the Alexander von Humboldt Award for Senior U.S. Scientists. He has a B.S. in physics from Stanford University and a Ph.D. in physics from Cornell University.

William Kalsbeek is professor of biostatistics and director of the Survey Research Unit at the University of North Carolina-Chapel Hill. His prior experience includes statistical research with the Office of Research and Methodology at the National Center for Health Statistics and at the Sampling Research and Design Center at the Research Triangle Institute. He is a fellow of the American Statistical Association and a member of the Biometrics Society and the American Public Health Association. His research interests and areas of expertise are in biostatistics, survey design and research, spinal cord injuries, and assessment. He has M.P.H. and Ph.D. degrees in biostatistics from the University of Michigan.

Jay Levinsohn is manager of technology issues at the Research Triangle Institute. Since joining RTI in 1974, he served as manager of information technology services from 1990 through 1998 and as senior research programmer/analyst since 1998. He is technical support manager for the National Household Survey on Drug Abuse, in which capacity he manages a technical support unit charged to assist a corps of 1,000 field staff running two shifts per day, seven days a week. Other projects he has directed include the Baseline Survey for Community Intervention Trial for Smoking Cessation, the Youth Attitude Tracking Study, and the Household Technology Survey. He has a B.S. in psychology and M.A. and Ph.D. degrees in quantitative psychology, all from the University of North Carolina.

Tony Manners is director of the research group in the Social Survey Division of the United Kingdom's Office for National Statistics, which is responsible for a range of continuous household surveys, evaluation research, and survey integration. He also has other projects like CAI documentation, including the TADEQ project, and development of web-CASI for official statistics. Until recently he was responsible for the survey contribution to UK National Statistics' program of harmonization of concepts and questions. He has worked in government social surveys for 27 years, including managing the Labour Force (LFS) and the Family Expenditure surveys during periods of extensive redevelopment of survey automation. He led the introduction of computer assisted interviewing in major UK social surveys, on the 1990 LFS. He has chaired two key users groups for Blaise: the International Blaise Users' Group (1992–2001); and the group of corporate license-holders which, in partnership with Statistics Netherlands, guides the strategic development of Blaise (since 1997, ongoing). He has an M.Phil. in social anthropology from London University.

Lawrence Markosian founded Reasoning, Inc., in 1984 and served as vice president for applications development from 1987 through 1995. From 1999 to 2001, he served as product manager for InstantQA, a tool for automated software defect detection that is Reasoning's principal product (InstantQA has since been renamed Illuma). During his tenure at Reasoning, he also served as director of training from 1996 through 1999. Since leaving Reasoning in 2001, he has served as senior technology transfer consultant at the NASA Ames Research Center, assisting in the development of survey software tools and engineering practices within NASA. He has a B.A. in mathematics from Brown University and was in the doctoral program in logic and philosophy at Stanford University.

Thomas J. McCabe, founder of McCabe & Associates in Columbia, Maryland, is a developer of software metrics and tools for software development, testing and maintenance. In 1998 he sold his business to venture capitalists and now heads McCabe Technologies. In 1999 he was appointed to lead the Washington, DC, chapter of the Chief Executive Officers (CEO) Club, a group consisting of founders and entrepreneurs. He also organized Mentors, a group of CEOs who have built, run, and sold companies and are in pursuit of their next challenge. He was the 1998 Stevens Award Lecturer on Software Development Methods. He has a B.S. from Providence College and an M.S. from the University of Connecticut, both in mathematics.

Martin Meyer joined Research Triangle Institute in 1998, where he is research programmer/analyst in the Research Computing Division. Currently, his main responsibilities are for software design and programming in support of the National Survey on Drug Use and Health, which is sponsored by the Substance Abuse and Mental Health Services Administration. Previously, he held a faculty position at Old Dominion University, where he taught and performed research in digital systems design. He has also worked and consulted for the U.S. Army Aeromedical Research Laboratory, the military's Joint Training Analysis and Simulation Center, the Army's Military Transportation Management Command, and the Virgnia Modeling Analysis and Simulation Center. He has B.S., M.S., and Ph.D. degrees in electrical and computer engineering from North Carolina State University.

Sarah Nusser is professor-in-charge of the Statistical Laboratory Survey Section and associate professor of statistics at Iowa State University. Her research interests include survey methods for welfare surveys, the use of statistics in biological and ecological studies, and computer-assisted survey information collection systems. Nusser has collaborated on a number of programs for the National Resources Conservation Service of the U.S. Department of Agriculture. She has a M.S. degree in botany from North Carolina State University and a Ph.D. in statistics from Iowa State University.

Thomas Piazza is the head of the Survey Documentation and Analysis (SDA) project in the Computer-Assisted Survey Methods Program at the University of California, Berkeley. He is also the senior survey statistician at the University's Survey Research Center and teaches applied sampling in the School of Public Health. He has a Ph.D. in sociology and many years of experience in designing, administering, documenting, and analyzing surveys.

Mark Pierzchala is senior analyst for computer-assisted interviewing methodology in the Washington, DC, office of Mathematica Policy Research, Inc. At the time of the Workshop on Survey Automation, he was senior systems analyst at Westat. Prior to joining Westat, he served as a mathematical statistician and a technical expert at the National Agricultural Statistics Service. His research interests include data editing of survey-collected data, development of complex CASIC instruments, specification of metadata for CASIC instrumentation and survey execution, usability for interviewers and data editors, and reduction of survey costs through integration and elimination of survey tasks. He has a masters degree in mathematical statistics from Michigan State University.

Jesse H. Poore is professor of computer science and software engineering at the University of Tennessee. In 2000, he was named director of the University of Tennessee-Oak Ridge National Laboratory Science Alliance. He conducts research in cleanroom software engineering and teaches software engineering courses. He has held academic appointments at Florida State University and Georgia Tech, has served as a National Science Foundation rotator, worked in the Executive Office of the President, and was executive director of the Committee on Science and Technology in the U.S. House of Representatives. He is a fellow of the American Association for the Advancement of Science. He has a Ph.D. in information in computer science from Georgia Tech.

Daryl Pregibon is division manager of the Statistics Research Department at AT&T Laboratories, a department responsible for developing a theoretical and computational foundation of statistics for very large data sets. Pregibon has nurtured interactions throughout AT&T, in fiber and microelectronics manufacturing, network reliability, customer satisfaction, fraud detection, targeted marketing, and regulatory statistics. His research contributions have changed from mathematical statistics to computational statistics and include such topics as expert systems for data analysis, data visualization, application-specific data structures for statistics, and large-scale data analysis. A fellow of the American Statistical Association, he received a masters degree in statistics from the University of Waterloo and a Ph.D. in statistics from the University of Toronto.

Harry Robinson leads software test productivity and model-based testing initiatives as part of Microsoft's Six Sigma Team. Prior to joining Microsoft in 1998, he was with Hewlett Packard for 3 years and with Bell Labs for 10 years. He teaches and consults on model-based testing within Microsoft and often speaks at industry conferences on test automation. He is currently creating a course on intelligent test automation for Microsoft Technical Education. A software developer for six years before switching to testing, he has a B.A. in religion from Dartmouth College and B.S. and M.S. degrees in electrical engineering from the Cooper Union for the Advancement of Science and Art.

Susan Schechter is senior statistician in the Statistical Policy Office of the U.S. Office of Management and Budget. She has spent her career working for the federal government, first at the Census Bureau and later at the Army Research Laboratory and the National Center for Health Statistics. Currently, she is the principal reviewer for all Census Bureau surveys and censuses at the U.S. Office of Management and Budget (OMB). Her OMB responsibilities also include working on implementation issues related to federal standards for data on race and ethnicity as well as measurement issues related to improving statistics on welfare, children, eduction, and health. She has a masters degree in human development research from Antioch University and completed undergraduate work at the University of Maryland, College Park.

Robert L. Smith is a former vice president of KnowledgeSet Corporation and is a former senior vice president of the Computer Curriculum Corporation (now NCS Learn, a division of NCS Pearson). He previously held an academic appointment in computer science at Rutgers University. He has a Ph.D. from Stanford University.

Roger Tourangeau is director of the Joint Program in Survey Methodology at the University of Maryland and a senior research scientist at the University of Michigan. He has been a survey researcher for more than 20 years. His research focuses on attitude and opinion measurement and on differences across methods of data collection; he also has extensive experience as an applied sampler. Tourangeau is well known for his work on the cognitive aspects of survey methodology and is the lead author of *The Psychology of Survey Responses*. Before joining JPSM and SRC, Tourangeau was a senior methodologist at the Gallup Organization, where he designed and selected samples and carried out methodological studies. Before that, he was at the National Opinion Research Center, where he founded and directed the Statistics and Methodology Center. There he carried out methodological studies and developed and executed sample designs for many federal and academic surveys. Named a fellow of the American Statistical Association in 1999, he has served on the editorial board of *Public Opinion Quarterly* and on the Census Joint Advisory Panel as member and alter chair of the ASA Subcommittee. Tourangeau has a Ph.D. in psychology from Yale University.